AQA Biology Revision Guide

A LEVEL YEAR 1 AND AS

David Applin

OXFORD UNIVERSITY PRESS

OXFORD
UNIVERSITY PRESS

Great Clarendon Street, Oxford, OX2 6DP, United Kingdom

Oxford University Press is a department of the University of Oxford.
It furthers the University's objective of excellence in research,
scholarship, and education by publishing worldwide. Oxford is a
registered trade mark of Oxford University Press in the UK and in
certain other countries

© David Applin

The moral rights of the authors have been asserted

First published in 2017

All rights reserved. No part of this publication may be reproduced,
stored in a retrieval system, or transmitted, in any form or by any
means, without the prior permission in writing of Oxford University
Press, or as expressly permitted by law, by licence or under terms agreed
with the appropriate reprographics rights organization. Enquiries
concerning reproduction outside the scope of the above should be sent
to the Rights Department, Oxford University Press,
at the address above.

You must not circulate this work in any other form and you must
impose this same condition on any acquirer

British Library Cataloguing in Publication Data
Data available

978-0-19-835178-8

10 9 8 7 6 5 4 3 2 1

Printed in Great Britain by Bell and Bain Ltd., Glasgow

Acknowledgements

Cover: Shutterstock/Michal Kowalski

Artwok by Q2A Media

LEICESTER LIBRARIES	
Askews & Holts	05-Apr-2017
	£9.99

AS/A Level course structure

This book has been written to support students studying for AQA AS Biology and for students in their first year of studying for AQA A Level Biology. It covers the AS sections from the specification, the content of which will also be examined at A Level. The sections covered are shown in the contents list, which also shows you the page numbers for the main topics within each section. There is also an index at the back to help you find what you are looking for. If you are studying for AS Biology, you will only need to know the content in the blue box.

AS exam

Year 1 content
1. Biological molecules
2. Cells
3. Organisms exchange substances with their environment
4. Genetic information, variation, and relationships between organisms

Year 2 content
5. Energy transfers in and between organisms
6. Organisms respond to changes in their internal and external environment
7. Genetics, populations, evolution, and ecosystems
8. The control of gene expression

A level exam

A Level exams will cover content from Year 1 and Year 2 and will be at a higher demand.

Contents

How to use this book — v

Section 1: Biological molecules — 2

1 Biological molecules — 2
1.1 Introduction to biological molecules — 2
1.2 Carbohydrates: monosaccharides — 4
1.3 Carbohydrates: disaccharides and polysaccharides — 5
1.4 Starch, glycogen, and cellulose — 6
1.5 Lipids — 7
1.6 Proteins — 8
1.7 Enzyme action — 10
1.8 Factors affecting enzyme action — 11
1.9 Enzyme inhibition — 13
Practice questions — 14

2 Nucleic acids — 15
2.1 Structure of RNA and DNA — 15
2.2 DNA replication — 17
2.3 Energy and ATP — 18
2.4 Water and its functions — 19
Practice questions — 20

Section 2: Cells — 21

3 Cell structure — 21
3.1 Methods of studying cells — 21
3.2 The electron microscope — 22
3.3 Microscope measurements and calculations — 23
3.4 Eukaryotic cell structure — 24
3.5 Cell specialisation and organisation — 25
3.6 Prokaryotic cells and viruses — 26
3.7 Mitosis — 27
3.8 The cell cycle — 28
Practice questions — 30

4 Transport across cell membranes — 31
4.1 Structure of the cell-surface membrane — 31
4.2 Diffusion — 32
4.3 Osmosis — 33
4.4 Active transport — 34
4.5 Co-transport and absorption of glucose in the ileum — 36
Practice questions — 37

5 Cell recognition and the immune system — 38
5.1 Defence mechanisms — 38
5.2 Phagocytosis — 39
5.3 T lymphocytes and cell-mediated immunity — 40
5.4 B lymphocytes and humoral immunity — 41
5.5 Antibodies — 42
5.6 Vaccination — 43
5.7 The human immunodeficiency virus (HIV) — 44
Practice questions — 45

Section 3: Organisms exchange substances with their environment — 46

6 Exchange — 46
6.1 Exchange between organisms and their environment — 46
6.2 Gas exchange in single-celled organisms and insects — 48
6.3 Gas exchange in fish — 49
6.4 Gas exchange in the leaf of a plant — 50
6.5 Limiting water loss — 51
6.6 Structure of the human gas-exchange system — 52
6.7 The mechanism of breathing — 54
6.8 Exchange of gases in the lungs — 55
6.9 Enzymes and digestion — 56
6.10 Absorption of the products of digestion — 58
Practice questions — 60

7 Mass transport — 61
7.1 Haemoglobin — 61
7.2 Transport of oxygen by haemoglobin — 62
7.3 Circulatory system of a mammal — 64
7.4 The structure of the heart — 65
7.5 The cardiac cycle — 66
7.6 Blood vessels and their functions — 67
7.7 Transport of water in the xylem — 68
7.8 Transport of organic molecules in the phloem — 68
7.9 Investigating transport in plants — 70
Practice questions — 71

Section 4: Genetic information, variation, and relationships between organisms — 72

8 DNA, genes, and protein synthesis — 72
8.1 Genes and the genetic code — 72
8.2 DNA and chromosomes — 73
8.3 Structures of ribonucleic acid — 74
8.4 Polypeptide synthesis: transcription and splicing — 75
8.5 Polypeptide synthesis: translation — 76
Practice questions — 78

9 Genetic diversity and adaption — 79
9.1 Gene mutation — 79
9.2 Meiosis and genetic variation — 80
9.3 Genetic diversity and adaptation — 82
9.4 Types of selection — 83
Practice questions — 84

10 Biodiversity — 85
10.1 Species and taxonomy — 85
10.2 Diversity within a community — 87
10.3 Species diversity and human activities — 89
10.4 Investigating diversity — 90
10.5 Quantitative investigations of variation — 91
Practice questions — 93

Answers to practice questions — 94
Answers to summary questions — 98

How to use this book

This book contains many different features. Each feature is designed to support and develop the skills you will need for your examinations, as well as foster and stimulate your interest in biology.

Worked example
Step-by-step worked solutions.

Common misconception
Common student misunderstandings clarified.

Question and model answers
Sample answers to exam style-style questions.

Summary questions

1. Distinguish between ionic bonds and covalent bonds.

2. Explain why polar molecules readily dissolve in water.

3. a Define the term 'macromolecule'.
   b How are macromolecules formed?

Specification references
→ At the beginning of each topic, there are specification references to allow you to monitor your progress.

Key term
Pulls out key terms for quick reference.

Synoptic link
These highlight how the sections relate to each other. Linking different areas of biology together becomes increasingly important, as many exam questions (particularly at A Level) will require you to bring together your knowledge from different areas.

Revision tip
Prompts to help you with your revision.

Practice questions at the end of each chapter including questions that cover practical and maths skills.

1.1 Introduction to biological molecules
Specification reference: 3.1.1

Molarity

One mole of an element is the amount of that element that contains 6.022×10^{23} atoms. This is the same number of atoms contained in 12 g of ^{12}C (carbon-12). 6.022×10^{23} is the Avogadro constant (or Avogadro number). Mol is the symbol for mole.

A molar solution (M) contains 1 mole of solute (substance which dissolves) in 1 dm³ of solution. To make up a molar solution of a compound, its molar mass should be calculated by adding together the molar mass of each of its component atoms.

> **Revision tip**
> Using the formula:
>
> molarity of solution = $\dfrac{\text{mass of substance}}{\text{molar mass of substance}}$
>
> the mass of substance needed to make up a solution of known molarity can be calculated.

> **Worked example**
>
> **Calculate the mass of sodium hydroxide (NaOH) needed to make 2.0 mol of sodium hydroxide solution.**
>
> The molar mass of each of the component atoms of sodium hydroxide is:
>
> sodium 23 g/mol
>
> oxygen 16 g/mol
>
> hydrogen 1 g/mol
>
> Adding together the molar mass of each of the atoms of sodium hydroxide then the molar mass of sodium hydroxide is 40 g/mol. So using the formula above left:
>
> $2.0 = \dfrac{\text{mass}}{40}$
>
> mass = $2.0 \times 40 = 80$ g
>
> You need 80 g of sodium hydroxide to make 2.0 mol of sodium hydroxide.

Chemical bonds

Atoms combine in different ways to form molecules of elements and compounds. The forces holding them together are chemical bonds.

- **Ionic bonds:** (groups of) atoms which carry an electric charge (+ve or –ve) are called ions. Ions carrying opposite charges attract one another. This electrostatic force is the ionic bond that holds the atoms together (an ionic compound).
- **Covalent bonds:** many compounds are not ionic and instead the atoms are held together by sharing a pair of electrons in their outer shells. The shared electron pair is the covalent bond which holds the atoms together (a molecule).
- **Hydrogen bonds:** a polar covalent bond forms where electrons are shared unequally between atoms. Because the molecule formed carries an uneven distribution of electric charge, it is called a polar molecule. An example of this is water as hydrogen bonds are polar covalent bonds.

Polarity means that water molecules can form hydrogen bonds with other water and polar molecules. Substances consisting of polar molecules are **hydrophilic**; those of non-polar molecules are **hydrophobic**.

> **Key terms**
>
> **Hydrophilic:** Readily dissolves in water.
>
> **Hydrophobic:** Does not readily dissolve in water.

Biological molecules

Condensation reactions and hydrolysis

Macromolecules are polymers. They are large and built up from monomers. These are smaller 'building block' molecules joined together. The categories of macromolecule and the monomers that form them are listed in Table 1.

▼ **Table 1** *Monomers form macromolecules*

Monomer		Category of macromolecule
monosaccharides (e.g. glucose, fructose)	→	polysaccharides (e.g. starch, cellulose)
fatty acid + glycerol	→	triglycerides
amino acids	→	peptides, polypeptides, proteins
nucleotides	→	nucleic acids (DNA, RNA)

Condensations join together monomers forming macromolecules. Each time monomers join together (e.g. amino acid with amino acid) a molecule of water is produced. The reverse (hydrolysis) occurs when the addition of water breaks down macromolecules into their component monomers.

$$n \text{ monomer molecules} \underset{\text{hydrolysis}}{\overset{\text{condensation}}{\rightleftharpoons}} \text{macromolecule} + n \text{ water molecules } (H_2O)$$

> **Revision tip**
> The different macromolecules that make up cells are often referred to as biological molecules. Metabolism refers to all of their reactions in cells.

> **Synoptic link**
> You can read more about monomers and macromolecules in Topics 1.2–1.6.

Summary questions

1. Distinguish between ionic bonds and covalent bonds. [1]

2. Explain why polar molecules readily dissolve in water. [1]

3. a Define the term 'macromolecule'. [2]
   b How are macromolecules formed? [1]

1.2 Carbohydrates: monosaccharides

Specification reference: 3.1.2

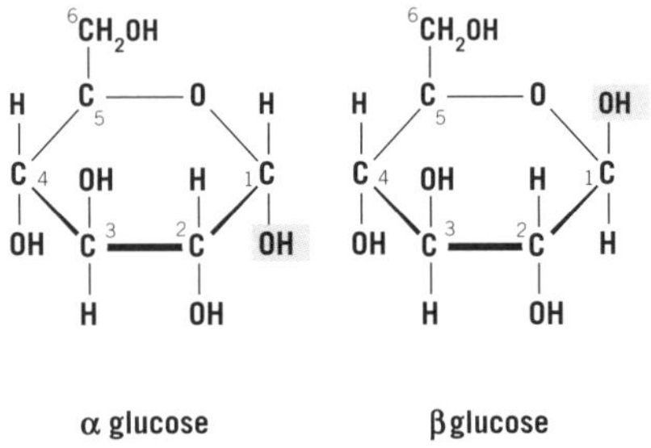

▲ Figure 1 *Isomers of glucose*

Categories of carbohydrate

Carbohydrates are compounds containing the elements carbon, hydrogen, and oxygen only. There are three main categories:

- **Monosaccharides:** single sugars, monomers of which carbohydrates are made (glucose, galactose, and fructose).
- **Disaccharides:** double sugars, one disaccharide molecule forms when two monosaccharide molecules combine (maltose, sucrose, and lactose).
- **Polysaccharides:** compounds that are formed from the combination of hundreds of monosaccharide molecules (starch, glycogen, and cellulose).

Monosaccharides

In general the molecular formula of monosaccharides is written as $(CH_2O)_n$. Some examples are given in Table 1. Notice that the ratio of hydrogen and oxygen atoms is 2:1, the same as water.

▼ Table 1 *Types of monosaccharide and molecular formulae*

Value of n	Type of monosaccharide	Molecular formula	Examples
3	Trioses	$C_3H_6O_3$	glyceraldehyde
5	Pentoses	$C_5H_{10}O_5$	ribose
6	Hexoses	$C_6H_{12}O_6$	glucose, fructose

In Table 1 glucose and fructose have the same molecular formula. However, their structural formulae show the different arrangement of the atoms of each molecule. They are structural isomers.

Glucose molecules also exist in different forms (isomers): (alpha) α-glucose and (beta) β-glucose (see Figure 1). If you imagine the molecules as 3D structures, with the hexagonal ring like a flat plate, the hydroxyl group (–OH) on carbon atom 1 is:

- below the plane of the ring in α-glucose
- above the plane of the ring in β-glucose.

Benedict's test for reducing sugars

All monosaccharides and some disaccharides are reducing sugars. They contain an aldehyde group (–CHO) or ketone group (>C=O) which reduces copper(II) ions (Cu^{2+}) to copper(I) ions (Cu^+) when heated in an alkaline solution. This is the basis of the Benedict's test for reducing sugars. Benedict's reagent turns from blue to red when heated with a reducing sugar.

Testing for glucose using colorimetry

Light is absorbed as it passes through a solution of a coloured substance. The amount of light absorbed by the solution depends on the concentration of the coloured substance. Benedict's test produces a range of colours depending on the concentration of reducing sugar in the solution being tested. The colorimeter can provide a quantitative result from this test.

The food is tested for glucose by boiling with Benedict's reagent. The resulting mixture is cooled and filtered. The absorbance of the supernatant (liquid) that remains after filtering is then measured in a colorimeter using a red filter. The greater the absorbance, the lower the glucose concentration. The absorbance can be compared with a calibration curve to give a quantifiable concentration of glucose.

Synoptic link

Read more about the disaccharide sucrose in Topic 1.3, Carbohydrates: disaccharides and polysaccharides.

Summary questions

1. The term 'carbohydrate' means hydrated carbon. Explain the link between the term and the general formula for a monosaccharide. [1]

2. Why are molecules of glucose and fructose described as structural isomers? [2]

3. Briefly describe a test that identifies the presence of non-reducing sugars. [4]

1.3 Carbohydrates: disaccharides and polysaccharides

Specification reference: 3.1.2

Disaccharides

A **disaccharide** is formed when two monosaccharide molecules combine.

- The reaction is a condensation. (The reverse hydrolysis breaks down maltose into its component α-glucose molecules.)
- The formation of a water molecule results in an oxygen 'bridge' joining the two molecules. The link is called a glycosidic bond.

In maltose, the glycosidic bond is between carbon atoms 1 and 4 of the adjacent α-glucose units. However, different disaccharides are produced if the combination of monosaccharide molecules is different and the links between carbon atoms in adjacent monosaccharide molecules is different.

For example:

- **sucrose** = α-glucose + β-fructose
- **lactose** = α-glucose + galactose

Testing for non-reducing sugars

The disaccharide sucrose is an example of a non-reducing sugar. It does not give a positive result to a simple Benedict's test. However, if sucrose solution is first heated with dilute hydrochloric acid then it hydrolyses into its component monosaccharide molecules. These then test positive when heated with Benedict's reagent. This is the basis of the test for a non-reducing sugar.

- The sugar solution gives a negative result when heated with Benedict's solution (no colour change).
- When hydrolysed (by heating with acid) and then neutralised (by adding sodium hydrogencarbonate), testing again gives a positive result to the Benedict's test. (The colour of the solution changes from blue to red.)

Polysaccharides

Polysaccharides are formed from many monosaccharides joined by condensation reactions. For example, starch is made from many condensation reactions forming glycosidic bonds between α-glucose molecules

Testing for starch

Starch doesn't dissolve in water but forms a suspension. Iodine doesn't dissolve in water either, but does dissolve in potassium iodide solution. When this iodine solution is added to the starch suspension, it turns from yellow-orange to an intense blue-black. This is the basis of the iodine test for starch. Starch turns iodine solution blue-black.

> **Revision tip**
> You can track the rate of digestion of starch using a colorimeter. A known volume of iodine is added to different concentrations of starch solution (standard volume), taking colorimeter readings for each concentration. Then you can plot colorimeter readings against concentration of starch solution to obtain a calibration curve. The unknown concentration of a starch solution can be determined from the calibration curve.

> **Revision tip**
> The general formula of polysaccharides is:
> $$nC_6H_{12}O_6 \rightarrow (C_6H_{10}O_5)_n$$ where $n = 10^2 - 10^3$

> **Summary questions**
>
> 1 Distinguish between condensation reactions and hydrolyses. [2]
>
> 2 Sucrose is a non-reducing sugar. Explain why it gives a positive result when heated with Benedict's reagent, following acid hydrolysis. [2]

1.4 Starch, glycogen, and cellulose
Specification reference: 3.1.2

Starch

Starch is a polysaccharide formed from condensation reactions between many molecules of α-glucose. It has two components: amylose and a mylopectin.

- **amylose:** consists of unbranched chains of α 1–4 glycosidic bonds between α-glucose molecules. Each chain coils like a spiral staircase into a helical structure.
- **amylopectin:** consists of branched chains with 1–6 glycosidic bonds between α-glucose molecules forming the branches.

The helical shape of amylose makes for a dense polysaccharide, enabling many subunits to be packed into a small space. Starch is an excellent storage molecule because it is not very soluble.

Glycogen

Glycogen is like amylopectin except that it consists of:

- fewer 1–4 glycosidic bonds between α glucose molecules
- many more 1–6 linkages.

The molecule is more branched than amylopectin, making glycogen less dense and slightly more soluble than starch.

Cellulose

Cellulose is a polysaccharide formed from condensation reactions between many molecules of β-glucose. Its fibres have high tensile strength.

- The –OH (hydroxyl) group on carbon atom 1 is above the plane of the ring of carbon atoms.
- The –OH group on carbon atom 4 is below the plane of the ring. As a result, for the –OH groups on carbon atoms 1 and 4 of adjacent β-glucose molecules to form a glycosidic bond, one molecule has to be flipped over (rotated through 180°) relative to its next-door neighbour.

The formation of β 1–4 glycosidic linkages produces rigid chain-like molecules.

- Hydrogen bonds cross-link chains into bundles called microfibrils.
- Further hydrogen bonding binds microfibrils into fibres.

Carbohydrates in living organisms

Food stores

Enzyme-catalysed hydrolyses break down starch and glycogen into glucose. When glucose is oxidised during cellular respiration energy is released, enabling cells to synthesise ATP. Condensation reactions polymerise glucose, forming molecules of starch and glycogen.

Starch and glycogen are almost insoluble which means that they have little effect on the osmotic properties of cells. Therefore, they can be stored in cells as energy reserves:

- starch is stored in plant cells
- glycogen is stored in animal cells.

> **Revision tip**
> High tensile strength means that each cellulose fibre is difficult to pull apart. 40% of the wall of a plant cell is made of cellulose. This means that plant cells can withstand large hydrostatic pressures inside them.

> **Summary questions**
>
> 1 How do the structures of α-glucose and β-glucose molecules differ? [2]
>
> 2 Briefly explain why plant cells do not burst when large hydrostatic pressures develop inside them as the result of osmosis. [3]
>
> 3 Why do starch and glycogen have little effect on the osmotic properties of cells? [4]

1.5 Lipids
Specification reference: 3.1.3

Lipids

Most lipids are mixtures of **triglycerides**.

- A triglyceride is an **ester** of fatty acids and glycerol.
- A molecule of fatty acid consists of a hydrocarbon chain and a carboxyl group (–COOH).

Molecules of fatty acid and glycerol combine to form a triglyceride. Note that:

- The combination of molecules of fatty acid with the molecule of glycerol is a condensation reaction.
- Three molecules of fatty acid are combined with one molecule of glycerol producing a triglyceride and three molecules of water.
  - A simple triglyceride is formed if the fatty acid molecules are the same.
  - A mixed triglyceride is formed if the fatty acid molecules are different.

The structure of triglycerides makes them useful in living organisms. They are:

- insoluble in water and store chemical energy, making fats and oils useful energy storage molecules
- readily hydrolysed to fatty acids and glycerol
- used in respiration in cells when glucose is in short supply
- can pass through the phospholipid bilayer of cell membranes.

Roles of lipids

Lipids have a range of functions in living things.

- **Energy source:** because they are insoluble in water lipids form long-term reserves of food (and energy) in plants and animals.
- **Insulation:** fat is a poor conductor of heat and electricity. A layer of fat beneath the skin can prevent heat loss (e.g. polar bears) and overheating (e.g. camels). The fatty myelin sheath surrounding neurones prevents voltage loss from axons.
- **Protection:** fat surrounds and protects delicate organs like the kidneys.
- **Structural:** phospholipids are an important component of cell membranes.
- **Molecules:** steroids and some hormones are lipids; e.g. oestrogen.

Phospholipids

Phospholipids are triglycerides that contain a phosphate group instead of one of the fatty acid components.

- The phosphate group dissolves in water and is therefore hydrophilic.
- The hydrocarbon chains of the two fatty acid components do not dissolve in water and are therefore hydrophobic.

The structure and function of cell membranes depend on the hydrophilic and hydrophobic properties of phospholipid molecules.

The emulsion test for lipids

Different tests may be used to detect the presence of lipids. One of them is the emulsion test:

- Mix ethanol and the test material in equal volumes.
- Shake to help dissolve any lipids present.
- Add an equal volume of water and shake the mixture again.
- A milky white emulsion indicates the presence of lipids.

> **Key term**
>
> **Lipids:** Compounds containing the elements carbon, hydrogen, and oxygen only.

> **Revision tip**
>
> Fats and oils produced by plants and animals are lipids. Fats are solid at room temperature; oils are liquid at room temperature.

> **Key term**
>
> **Ester:** A substance formed from the reaction between an acid and an alcohol.

> **Revision tip**
>
> The R-group of a fatty acid refers to the number of carbon atoms and the number of hydrogen atoms bonded to each carbon atom. The R-group of a fatty acid may be saturated or unsaturated.

> **Summary questions**
>
> 1 Distinguish between saturated and unsaturated fatty acids. [2]
>
> 2 What is an emulsion? [2]

1.6 Proteins

Specification reference: 3.1.4.1

> **Key term**
>
> **Proteins:** Compounds containing the elements carbon, hydrogen, oxygen, and nitrogen. Some also contain sulfur.

> **Key term**
>
> **Amino acids:** The monomers (the building blocks) from which proteins are made. There are 20 different amino acids.

> **Key term**
>
> **Dipeptide:** Formed when two amino acid molecules combine.

Amino acids: the building blocks of proteins

Each amino acid has a different R group (e.g. if R is a hydrogen atom then the amino acid is glycine). There are 20 different R groups, which is why there are 20 different amino acids.

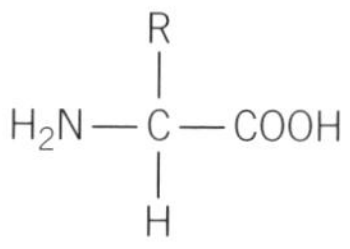

▲ **Figure 1** *The general structure of amino acid*

Peptide bonds

The carboxyl group of one amino acid molecule reacts with the amino group of the other amino acid molecule; a molecule of water is eliminated. The reaction is a condensation; the reverse hydrolysis would break the dipeptide into its component amino acid molecules. As a result a peptide bond forms between the two amino acid molecules.

The type of dipeptide formed depends on the structures of the R groups. The more amino acid units joined together by condensation reactions, the larger is the resulting polymer.

Protein structure

Primary (1°) structure

The order in which one amino acid unit follows another in the polypeptide chain(s) is unique to each protein. This unique amino acid sequence dictates the:

- structure
- chemical properties
- function

of the particular protein. The sequence of amino acid units is called the primary (1°) structure of the protein.

Secondary, tertiary, and quaternary structures of proteins arise from the primary structure of polypeptides.

The primary structure of a polypeptide chain allows:

- hydrogen bonds to form between different amino acids along the chain
- interactions between R groups of the amino acids along the chain.

As a result the polypeptide chain bends and twists, giving rise to the secondary, tertiary, and quaternary structures that shape the protein molecule.

Secondary (2°) structure

Secondary structure arises because of hydrogen bonding between the oxygen of the >C=O group of one amino acid unit and the hydrogen of the >NH group of another amino acid unit. If this bonding occurs within one polypeptide chain, the chain coils into an alpha helix (α-helix). If this bonding occurs between different, parallel polypeptide chains, the chains fold into a beta pleated sheet (β-pleated sheet).

Go further

Collagen is a protein with a quaternary structure. It is an important component of ligaments and tendons.

a Explain why collagen is described as having a quaternary structure.

b Distinguish between ligaments and tendons.

c How does the structure of a molecule of collagen adapt (suit) collagen to its function?

Tertiary (3°) structure

Tertiary structure arises when the α-helices and β-pleated sheets of many proteins fold and coil into a shape, which is held in place by weak and strong covalent chemical bonds between different groups in the polypeptide chain.

Disulfide bonds (S–S) between units of the amino acid cysteine are strong bonds. Weak bonds include:

- **hydrogen bonds** between the >NH group and >C=O group of different amino acids
- **ionic interactions** from attractions between positively and negatively charged areas of the molecule
- **van der Waals** forces between non-polar R-groups of the molecule.

Weak bonds mean that the tertiary structure of proteins is flexible and can change in response to changes in pH and temperature, for example. The changes are reversible and enable proteins to carry out their many different functions.

Quaternary (4°) structure

Quaternary structure arises when a protein molecule consists of two or more polypeptide chains. Haemoglobin (the oxygen carrying pigment in red blood cells) is an example. Other examples include enzymes and antibodies. The way in which the chains fit together is maintained by the same types of chemical bond that hold together its tertiary structures.

Testing for protein

Adding an alkaline solution of copper(II) sulfate to material that contains peptides or proteins produces a pink to purple colour. The test is called the biuret test. It detects peptide bonds. All peptides and proteins, therefore, give a positive result. The biuret test is a qualitative test but can also be used semi-quantitatively. The colours produced depend on the number of peptide bonds in the test substance. Proteins give a characteristic purple colour with the biuret test because they contain more peptide bonds per molecule compared with peptides and polypeptides.

Summary questions
1 Which elements make up a molecule of amino acid? [2]
2 Briefly describe the biuret test for proteins. [3]
3 Explain the differences between the secondary structure and tertiary structure of proteins. [3]

1.7 Enzyme action
Specification reference: 3.1.4.2

Enzyme action

Enzymes are catalysts. In general, a catalyst:

- alters the rate of a chemical reaction
- is effective in small amounts
- is involved in but unchanged by the chemical reaction it catalyses.

All the features of catalysts are also features of enzymes. In addition enzymes are:

- specific in their action, catalysing a particular chemical reaction or type of reaction
- sensitive to changes in pH and temperature.

Only a small part of an enzyme molecule binds with its substrate molecule. This part is called the active site. It consists of just a few of the amino acid units that make up the enzyme molecule as a whole.

Enzyme action may be intracellular, catalysing reactions inside the cell where the enzyme is made. Cellular respiration is an example. Different enzymes catalyse the oxidation of substrates (e.g. glucose), releasing energy, which the cell uses to synthesise ATP.

The action of other enzymes may be extracellular. They catalyse reactions outside of cells. For example, amylase made in cells of the salivary glands and pancreas catalyses the digestion of starch in the mouth and duodenum of the small intestine respectively.

> **Revision tip**
> The substance an enzyme helps to react is the substrate. The substance formed by the reaction is the product. When an enzyme and its substrate bind together an enzyme–substrate complex is formed. Within the enzyme–substrate complex, the substrate undergoes reaction forming an enzyme–product complex. The product(s) then leave the active site. The enzyme can now catalyse another reaction.

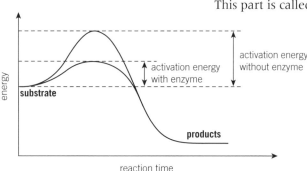

▲ Figure 1 Enzymes lower activation energy

Enzymes lower activation energy

Activation energy is the amount of energy required to bring about any particular chemical reaction. Without it chemical reactions will not take place. Enzymes lower activation energy by forming enzyme–substrate complexes. They enable reactions which would need high temperatures in the laboratory to take place at body temperature.

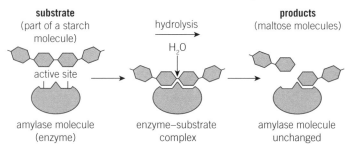

▲ Figure 2 The enzyme amylase binds with starch, catalysing the breaking of alternate glycosidic bonds. The reactions are hydrolyses and maltose is formed

How do enzymes work?

The shape of an enzyme, like all proteins, is the result of its tertiary structure and its active site has a precise shape. An enzyme will bind with a particular substrate molecule because the shape of the active site is complementary to (opposite to) the shape of the substrate molecule. The two shapes fit like a key fits into a lock. The idea of lock and key helps to explain why we say that an enzyme is specific to its particular substrate. Only the shape of the substrate molecule in question fits the active site of the enzyme. Figure 2 shows an example of this.

'Lock and key' suggests that the active site of an enzyme and its substrate are exactly complementary. Recent work favours the induced-fit hypothesis. The active site and substrate are fully complementary only after binding has taken place. The initial binding of a substrate molecule to the active site alters the shape (tertiary structure) of the active site. As a result the shape of the substrate molecule then alters, assisting the reaction to take place.

> **Summary questions**
>
> 1 What is activation energy and how do enzymes affect it? [2]
>
> 2 Use the induced-fit hypothesis to describe the binding of an enzyme with its substrate. [3]

1.8 Factors affecting enzyme action
Specification reference: 3.1.4.2

Measuring rate of an enzyme-controlled reaction

The time taken for the appearance of product / disappearance of substrate is a measure of the rate of an enzyme-controlled reaction. The appearance of product / disappearance of substrate can be monitored by a change in colour of the reaction mixture, e.g. the enzyme trypsin catalyses the breakdown of protein into peptides. The rate of reaction can be measured as the time taken for the reaction mixture to change.

$$\text{protein} \xrightarrow{\text{trypsin}} \text{peptides}$$
$$\text{white} \qquad\qquad \text{colourless}$$

A change in amber-coloured potassium iodide solution to blue-black when added to a liquid indicates the presence of starch in the liquid. The enzyme amylase catalyses the breakdown of starch to maltose. The rate of reaction can be measured as the time taken for the reaction mixture to change

$$\text{starch} \xrightarrow{\text{amylase}} \text{maltose}$$
$$\text{blue-black} \qquad \text{amber}$$

when potassium iodide solution is added to the mixture.

Factors affecting the activity of enzymes

Temperature

In general, the rate of chemical reactions increases with temperature (doubles for every 10°C rise in temperature). But for enzyme-catalysed reactions, this is true only within a limited range of temperature. Once an enzyme reaches its optimum, any further increase in temperature causes a decrease in the rate of reaction. The decrease is caused by a permanent change in the shape of the enzyme (and its active site). We say that the enzyme is denatured. Its active site is no longer complementary with the substrate molecule.

> **Revision tip**
>
> $$\text{Rate} = \frac{\text{change}}{\text{time}}$$

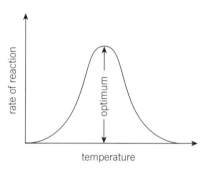

▲ **Figure 1** *Enzyme activity varies with temperature*

> **Common misconception: Optimum temperature**
>
> Because the body temperature of most mammals (including us) is normally constant at 37°C, we think this temperature is the optimum for all enzymes in all species. However enzyme activity of fish in Arctic / Antarctic waters is optimal at −2°C, and the enzymes denature if water temperature increases to 5°C. Different species of bacteria flourish in hot springs and deep sea hydrothermal vents where water may approach boiling point (100°C). Activity of their enzymes is optimal at these temperatures. The term thermophiles refers to the bacteria. We make use of heat-stable enzymes extracted from thermophilic bacteria in washing powders / liquids. The examples highlight the fact that the optimal temperature for enzyme activity varies depending on species and the environment in which species live.

Biological molecules

> **Question and model answer**
>
> **Q** What does 'optimum' mean when we talk about the effect of temperature on enzyme activity?
>
> **A** *The 'optimum' is the temperature value at which the number of collisions between the active site of an enzyme and its substrate is at a maximum. As a result the rate of the enzyme-catalysed reaction is at a maximum. As the temperature value decreases below the optimum, then the number of collisions between the enzyme's active site and its substrate reduces, reducing the rate of the enzyme-catalysed reaction. The enzyme is increasingly deactivated.*

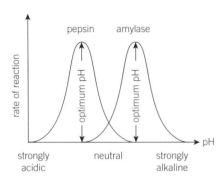

▲ **Figure 2** *Enzyme activity varies with pH*

pH

A change of pH from the optimum for a particular enzyme alters the electric charge carried by the amino acid units forming the active site. The enzyme (and its active site) is denatured, and is no longer complementary with the substrate molecule. This causes a decrease in the rate of the enzyme-catalysed reaction.

Concentration of enzyme

Enzymes are not used up during catalysis, and can be used over and over again. They therefore work very well at low concentrations. Increasing the enzyme concentration provides more active sites so the rate of enzyme activity increases as long as the substrate is present in excess (more than enough).

Concentration of substrate

An increase in the concentration of substrate affects the rate of reaction for a fixed concentration of enzyme.

- If there is an excess of enzyme, the rate of reaction is directly proportional to the concentration of the substrate (1 - see Figure 3).
- When all the enzyme active sites are occupied by substrate molecules the rate of reaction is limited and becomes constant (2 - see Figure 3).

The effects of substrate concentration on the rate of reaction assume that all the other conditions that affect the rate of enzyme-catalysed reactions (e.g. pH, temperature) are constant.

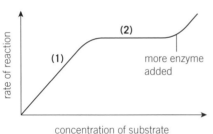

▲ **Figure 3** *Affect of substrate concentration on the rate of enzyme catalysed reaction. If more enzyme is added to the mixture when the rate of reaction is constant then the increase in the rate of reaction is proportional to the concentration of the substrate*

> **Summary questions**
>
> 1. Explain why the rate of an enzyme-catalysed reaction increases as temperature increases to the optimum for the reaction in question. [2]
>
> 2. For a fixed concentration of substrate, explain why the rate of an enzyme-catalysed reaction first increases with the addition of enzyme but is then limited with the addition of even more enzyme. [4]
>
> 3. Explain the differences between the denaturation and deactivation of an enzyme. [2]

1.9 Enzyme inhibition
Specification reference: 3.1.4.2

Competitive inhibition
A competitive inhibitor is a substance that combines with the active site of an enzyme, preventing its normal substrate from binding with it (see Figure 1).

Non-competitive inhibition
A **non-competitive inhibitor** is a substance that combines with some part of an enzyme molecule other than its active site. The change in the shape of the enzyme molecule causes a change in the shape of its active site. The substrate molecule is no longer able to bind to the active site.

The inhibition may be:

- **Reversible:** breaking up the inhibitor–enzyme complex is possible.
- **Irreversible:** breaking up the inhibitor–enzyme complex is not possible. Heavy metals such as arsenic (As) and mercury (Hg) are irreversible non-competitive inhibitors, which is why the substances are poisonous.

> **Key term**
>
> **Inhibitors:** Substances that reduce the rate of reaction when added to an enzyme/substrate mixture.

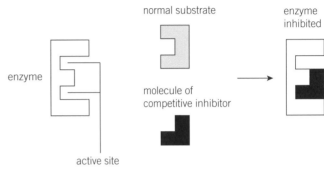

▲ **Figure 1** *Enzyme inhibition*

> **Revision tip**
> A metabolic pathway is a sequence (particular order) of reactions where a particular molecule is converted into another different one by way of a series of intermediate compounds.
>
> intermediate compounds
>
> molecule A → molecule B → molecule C → molecule D → molecule E

If the product of one reaction of a metabolic pathway inhibits one of the enzymes preceding its formation, the product will act as an inhibitor of the whole pathway. The process is an example of non-competitive inhibition called end-point inhibition.

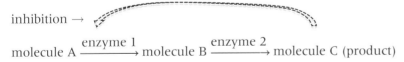

inhibition →

molecule A $\xrightarrow{\text{enzyme 1}}$ molecule B $\xrightarrow{\text{enzyme 2}}$ molecule C (product)

For example molecule C inhibits enzyme 1. Less of molecule C is produced so inhibition of enzyme 1 decreases. More of molecule C is produced and inhibition of enzyme 1 increases. End-point inhibition is an example of negative feedback; the mechanism of homeostasis. It controls the rates of reaction of metabolic pathways.

Allosteric inhibition
In some enzymes, a chemical group other than the active site can bind with a substance other than the normal substrate. The group is called the allosteric site. An **allosteric inhibitor** is a substance that reversibly combines with the allosteric site.

> **Summary questions**
>
> 1. Distinguish between competitive inhibition and non-competitive inhibition. [2]
>
> 2. Explain why heavy metals are poisonous substances. [1]
>
> 3. How does end-point inhibition control the rate of reactions of a metabolic pathway? [1]

Chapter 1 Practice questions

1. A teacher described a water molecule in terms of a hydrogen atom and two oxygen atoms held together by covalent bonds.

   a. What is a covalent bond? *(1 mark)*

   b. Why did the teacher also describe water as a polar molecule? *(2 marks)*

   c. Give *three* reasons why the polarity of water molecules enables living things to exist. *(3 marks)*

2. Many biological molecules are made up of many molecules of monomers joined together by chemical bonds.

   a. Name the type of chemical reaction that produces polymers from joining together monomers. *(1 mark)*

   b. Table 1 identifies categories of biological molecules. Name the different monomers and the chemical bonds that join them together, producing the biological molecules listed.

   ▼ Table 1

Category of biological molecules	Monomers	Chemical bond
Triglyceride		
Protein		
Carbohydrate		

   (7 marks)

3. The graph shows how temperature affects the rate of an enzyme-controlled reaction:

   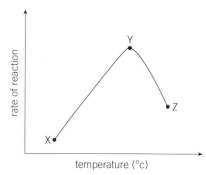

   a. Explain the change in the rate of reaction between:

      i. X and Y *(4 marks)*

      ii. Y and Z. *(3 marks)*

   b. How might high temperature affect the active site of the enzyme? *(3 marks)*

4. The biuret test detects peptide bonds. The colours produced depend on the number of peptide bonds in the test substance. Describe how the biuret test can be used semi-quantitatively to determine the relative concentrations of different protein solutions.

   (3 marks)

2.1 Structure of RNA and DNA
Specification reference: 3.1.5.1

Nucleotides

Nucleotides are monomers consisting of three components; a molecule each of:

- a pentose sugar (deoxyribose (DNA), ribose (RNA))
- a nitrogen-containing organic base (either adenine (A), thymine (T), guanine (G), or cytosine (C) in DNA; uracil (U) substitutes thymine in RNA
- a phosphate group.

Condensation reactions join the components together forming a nucleotide.

Comparing nucleic acids

Table 1 compares nucleic acids, summarising their similarities and differences.

▼ **Table 1** *Nucleic acids compared. Note that the base uracil (U) substitutes for thymine in mRNA and tRNA*

	DNA	mRNA	tRNA
Polynucleotide strand	double	single	single
Synthesis	replicated by parent DNA	transcribed against DNA template	transcribed against DNA template
Number of bases	variable: many thousands	variable: many hundreds to a few thousands	75–90
Pentose sugar	deoxyribose	ribose	ribose
Bases	ATCG	AUCG	AUCG
Ratio of bases	A:T = 1 C:G = 1	variable	variable
Shape of molecule	double helix	single strand in different shapes	single strand in the form of a 'clover leaf'
Location	mainly in the nucleus (also mitochondria and chloroplasts)	made in the nucleus; found in the cytoplasm	made in the nucleus; found in the cytoplasm
Function	carries genetic information inherited from parent DNA	carries genetic information from the nucleus to the ribosomes where polypeptides are synthesised	binds to amino acids and carries them to the mRNA combined with ribosomes

> **Key term**
>
> **Genetic:** The inheritance of DNA by daughter cells when a parent cell divides.

> **Synoptic link**
>
> The importance of DNA and RNA as information-carrying molecules is covered in Topic 8.4, Polypeptide synthesis: transcription and splicing.

> **Synoptic link**
>
> There are different types of RNA (see Table 1). You can read more about RNA in Topic 8.3, Structures of ribonucleic acid.

DNA structure

Many condensation reactions join together deoxyribose units forming a strand of DNA. The strand is a polynucleotide (many nucleotides joined together):

- adjacent sugar rings link the phosphate group from carbon atom 3 of one sugar to carbon atom 5 of the next sugar in line
- the links are **phosphodiester** bonds and hold the DNA strand together.

A molecule of DNA is made of two polynucleotide strands. The strands coil round each other forming a 'double helix'. This is stabilised by hydrogen bonds between the bases attached to the two strands:

- A only bonds with T
- G only bonds with C.

This arrangement is called complementary base pairing:

- one strand runs from carbon atom 3′ → 5′ → 3 in one direction
- its partner strand runs 5′ → 3′ → 5′ in the opposite direction.

We say that the strands are anti-parallel.

RNA structure

RNA is also a polynucleotide:

- its sugar units are not deoxyribose but ribose (carbon number 2 has an −OH group)
- the base thymine is replaced by the base uracil.

Summary questions

1. What does the term 'genetic' mean? [1]
2. Briefly explain the relationship between phosphodiester bonds, complementary base pairing, and a molecule of double-stranded DNA. [3]
3. What are the components of a nucleotide? [2]
4. Explain the role of each type of RNA: mRNA, tRNA, and rRNA. [4]

2.2 DNA replication
Specification reference: 3.1.5.2

Making an exact copy

During replication DNA makes an exact copy of itself. The process is part of the division of a cell nucleus.

There are three stages.

1. Unwinding and unzipping the original DNA molecule.
   - DNA helicase enzymes catalyse the:
     - unwinding of double-stranded DNA
     - breaking of hydrogen bonds linking the base pairs of the two strands of DNA.
   - As a result the double helix unzips as the base pairs separate.

2. Building the new polynucleotide chain of DNA.
   - Nucleotides free in solution within the nucleus each link with their complementary base on either of the unzipped strands of DNA (each called a template strand). Linkage is catalysed by the enzyme DNA polymerase.
   - The process repeats itself again and again in the $5' \rightarrow 3' \rightarrow 5'$ direction against each template strand.
   - As a result a new polynucleotide strand grows against each template strand.

3. Separation of the new DNA molecules.
   - When all of the bases of each template strand of DNA are each joined with the complementary base of a free nucleotide and these nucleotides have linked together, replication is complete.
   - The two new molecules of DNA separate.

Semi-conservative replication

DNA replication occurs during the s-phase of interphase. The process is semi-conservative in that each new DNA molecule consists of a strand of template DNA (arising from the unzipping of the original double helix) and a strand of DNA formed as a complement of the template strand. Genetic continuity from one generation of cells to the next (the cells are genetically identical) is ensured through the semi-conservative replication of DNA.

> **Revision tip**
> The replication of DNA is a very accurate process. Errors of copying (gene mutations) occur about once for every billion (1×10^9) nucleotides linked to template DNA. It seems that DNA polymerase 'proofreads' its own activity of linking nucleotides and cuts out any linkage 'mistakes' from the growing strand of new DNA.

> **Revision tip**
> Helicases are a category of enzymes that catalyse the processes that separate the strands of double-stranded DNA molecules. Their activity depends on the energy released from the hydrolysis of ATP; ATP $\rightarrow$ ADP + P_i.

Summary questions

1. What are the roles of DNA helicase and DNA polymerase during DNA replication? [2]
2. Why is DNA replication described as semi-conservative? [2]
3. What does the phrase 'DNA template strand' mean? [2]

2.3 Energy and ATP
Specification reference: 3.1.6

> **Revision tip**
> Energy is not a thing in itself. It refers to the changes in state (e.g. molecules change from one type to another type) that maintain life.

Energy and ATP

ATP is short for adenosine triphosphate. Its molecule consists of the sugar ribose to which is attached the base adenine and a chain of three phosphate groups. The combination of ribose and adenine forms the adenosine part of the molecule.

ATP is formed when a phosphate group (P_i) binds to a molecule of ADP (adenosine diphosphate).

The reaction is:

- a phosphorylation: a type of reaction where a phosphate group is added to another molecule
- endothermic: a type of reaction which absorbs energy
- catalysed by the enzyme ATP synthase.

The energy driving the synthesis of ATP comes from:

- light in the light-dependent reactions of photosynthesis
- sugars (also lipids and proteins) which are oxidised in the reactions of cellular respiration.

In each case the energy released drives the phosphorylation of ADP, and is stored in the chemical bonds of the ATP molecules produced.

Releasing energy from ATP

ATP is an immediate source of energy for biological processes. It is very soluble in water and often described as the universal energy currency found in the cells of all living organisms, from bacteria, to oak trees, to humans. When a molecule of ATP combines with a water molecule, the bond binding the endmost phosphate group to the rest of the ATP is broken. The reaction is hydrolysis. During the reaction energy is consumed breaking the bond but more energy is released as other bonds are formed. Overall the reaction is therefore exothermic.

The hydrolysis of ATP is catalysed by the enzyme ATP hydrolase. The energy released is coupled to, and drives, the processes that maintain life. The processes include.

- **Anabolic reactions** which result in the synthesis of polymers from building block units, for example monosaccharides → polysaccharides.
- **Active transport** of substances against their concentration gradients across cell membranes, for example:
   - transferring glucose from blood to liver cells
   - the exchange of sodium ions (Na^+) and potassium ions (K^+) across the membrane of the axon of a nerve cell, generating an action potential.
- **Muscle contraction** during which muscle fibres shorten.

A proportion of the energy from the hydrolysis of ATP is released as heat energy. In birds and mammals some of the heat helps to maintain a constant body temperature. The rest is transferred to the environment.

> **Summary questions**
>
> 1. Name two sources of energy that drive the synthesis of ATP. [2]
>
> 2. Why is adenosine triphosphate so named? [2]
>
> 3. Explain why the synthesis of ATP is described as an endothermic reaction. [2]

2.4 Water and its functions
Specification reference: 3.1.7, 3.1.8

Dipolarity and hydrogen bonding
The shape of a water molecule is asymmetrical:

- the molecule has a dipole:
  - the oxygen atom carries a small negative charge
  - the hydrogen atoms each carry a small positive charge
- negative and positive charges balance
- as a result a water molecule has no *net* charge.

The solvent properties of water
Ions and polar molecules readily dissolve in water because of the polarity of their molecules. The liquid in which they dissolve is called the solvent. Solutes (which dissolve in water) are said to be hydrophilic because they readily form hydrogen bonds with water molecules. Because of its solvent properties, water is a metabolite (substance taking part in the chemical reactions in cells) in many metabolic reactions (the chemical reactions in cells), including hydrolysis and condensation reactions. Non-polar molecules are hydrophobic and do not dissolve in water (they are insoluble).

> **Synoptic link**
> Xylem vessels will be covered in Topic 7.7, Transport of water in the xylem, and condensation and hydrolysis reactions were covered in Topic 1.3, Carbohydrates: disaccharides and polysaccharides.

The heat capacity of water
The temperature of a substance is a measure of the motion of its molecules. The *more* heat absorbed, the greater the molecular motion, and so the temperature of the substance increases. Water can absorb more heat than most fluids before its temperature increases noticeably. In other words water has a high heat capacity. This helps to damp down large swings in temperature, providing a stable environment for chemical reactions in cells and for organisms living in water.

Cohesion and surface tension
Cohesion means the capacity of something to resist pulling apart under tension. Water has high cohesion because its molecules are joined by hydrogen bonds. These cohesive forces can also act on the solution which moves through the xylem of plants. As a result unbroken columns of water move through the plant from the roots to the leaves.

Vaporisation
Hydrogen bonding between water molecules hinders the molecules from escaping as vapour. Therefore a large input of heat energy is needed to break the hydrogen bonds, and for them to remain broken. Only then does the molecular motion of the water molecules increase with the result that the molecules can escape into the air (vaporisation). The escaping molecules carry away heat energy and the water's surface temperature decreases. This is why sweat helps the body to cool down on hot days.

> **Revision tip**
> Without water life as we know it would not exist. For example 60% of the human body consists of water.

> **Revision tip**
> The hydrogen–oxygen attractions holding the molecules together form hydrogen bonds.

> **Revision tip**
> Sweat is about 99% water.

> **Synoptic link**
> An ion is an electrically charged (+ve or –ve) atom or group of atoms. Different ions are important components of biological molecules, for example iron ions in haemoglobin bind with oxygen molecules.

> **Summary questions**
>
> 1 a Explain the meaning of the phrase 'a water molecule is a dipole'. [3]
>   b Why is the dipolarity of water fundamental to the existence of life? [2]
>
> 2 Distinguish between heat, temperature, and heat capacity. What is the significance of the high heat capacity of water for aquatic organisms? [4]
>
> 3 Explain the significance of the cohesive forces between water molecules for plants. [1]

Chapter 2 Practice questions

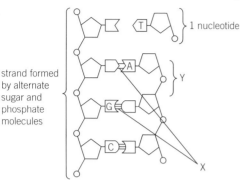

▲ Figure 1

1. Figure 1 represents part of a DNA molecule:

   The letters represent the bases adenine (A), thymine (T), cytosine (C), and guanine (G).

   a Complete the empty boxes with the letter of the correct base. *(2 marks)*

   b Name the chemical bonds X and Y. *(2 marks)*

   c Which of the chemical bonds break when a DNA molecule replicates? *(1 mark)*

   d Name the enzyme which catalyses the breaking of the chemical bonds referred to in **c**. *(1 mark)*

   e At what stage does DNA replicate in the cell cycle? *(1 mark)*

2. When a DNA molecule separates, each of its strands is a template against which a new strand forms. The two new molecules are identical to each other and the original molecule.

   a Explain why DNA replication is described as semi-conservative. *(2 marks)*

   b During replication, new strands of DNA form against their respective template strands in opposite directions. Explain why. *(2 marks)*

   c Explain why replicated DNA molecules are identical to each other and the original molecule *(2 marks)*

   d The mass of guanine of a sample of DNA was measured as 22 arbitrary units out of 100 arbitrary units in total:

   i calculate the mass of thymine in the sample *(2 marks)*

   ii explain how you arrived at your answer in **i**. *(5 marks)*

3. Figure 2 represents the structural formula of a molecule of ATP:

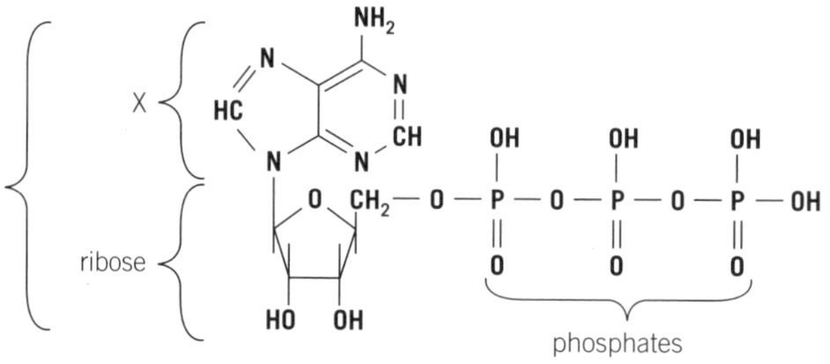

   ▲ Figure 2

   a Name the components of the molecule labelled X and Y. *(2 marks)*

   b Using the diagram, explain why the sugar component is identified as ribose and not deoxyribose. *(2 marks)*

   ATP is produced by the combination of ADP and a phosphate group (P_i). More energy is released during the reaction than is absorbed.

   c Name the type of reaction where a phosphate group is added to another substance. *(1 mark)*

   d Give the term which describes a reaction that absorbs more energy than it releases. *(1 mark)*

   e List the sources of energy that enable plant cells to synthesise ATP. *(2 marks)*

   f Identify two biological processes that require ATP. *(2 marks)*

3.1 Methods of studying cells
Specification reference: 3.2.1.3

Seeing cells

Most cells are too small to be seen with the unaided eye. Different types of microscope help us look at cells.

The transmission electron microscope (TEM) and scanning electron microscope (SEM) reveal the structure of cells in more detail than the optical (light) microscope. The level of detail seen (resolution of the image) is a measure of the resolving power of the microscope.

▼ **Table 1** *Wavelength and resolution*

Property	Light microscope	TEM	SEM
wavelength	400–700 nm	0.005 nm	0.005 nm
resolving power	200 nm	0.5 nm (technical difficulties reduce resolution of the image compared with the theoretical best value)	3–10 nm (less than TEM)
maximum useful magnification	× 2000	× 2,000,000	× 200,000

Fractionation and ultracentrifugation

Fractionation (breaking up cells suspended in a buffered liquid) releases the structures within cells. Ultracentrifugation involves spinning test tubes, each containing a suspension of organelles (cell structures), in a centrifuge at high speed. This separates the different organelles. All stages take place in chilled conditions to reduce the self-digestion of organelles (reactions catalysed by the release of enzymes because of fractionation). The different techniques of ultracentrifugation provide pure samples of organelles. Each type of organelle can then be analysed separately. This means that the outcome of analysis will not be affected by other types of organelle.

> **Key term**
>
> **Resolving power of a microscope:** The ability to distinguish between structures lying close together.

> **Revision tip**
>
> Magnifying structures to the limits of a microscope's resolving power reveals more and more detail. But structures lying closer together than the ability of the microscope to resolve them appear as a single structure. As a result, magnifying structures beyond a microscope's resolving power enlarges the image but does not improve clarity of detail. The limit of resolution is one half the wavelength of the electromagnetic radiation (light or electrons) used to illuminate the specimen under observation.

> **Key term**
>
> **Magnification:** The number of times larger the image of an object is than the object's actual size.

Summary questions

1. Explain the difference between the magnifying power and the resolving power of a microscope. [2]

2. Why is the resolving power of a transmission electron microscope greater than that of an optical microscope? [1]

3. Explain why centrifugation can provide pure samples of organelles. [2]

3.2 The electron microscope

Specification reference: 3.2.1.3

The transmission electron microscope (TEM)

In the TEM a beam of electrons passes through the specimen under observation. More electrons pass through some parts of the specimen than others. The electrons then hit a screen coated with phosphorescent material that glows on their impact. The more electron hits, the brighter the glow. The image produced is a highly detailed shadow of the specimen.

The scanning electron microscope (SEM)

In the SEM a beam of electrons *scans* the specimen and knocks electrons loose from the specimen's surface. These electrons are captured and a computer processes the information, assembling a detailed three-dimensional image of the specimen's surface features.

> **Revision tip**
> The maximum magnification of the SEM is one order of magnitude ($\times 10$) less than the TEM.

Advantages and limitations

High magnification without loss of resolution (the detailed structures of specimens remain clear) is an important advantage of the different types of electron microscope compared with optical microscopes.

However, there are limitations as well. Tables 1 and 2 compare advantages and limitations.

▼ **Table 1** *Disadvantages and advantages*

Disadvantages: optical microscope	Advantages: TEM
Maximum useful magnification × 2000	Maximum useful magnification × 2,000,000
Resolves structures up to 200 nm apart	Resolves structures up to 0.5 nm apart
The sections of tissues are relatively thick. Therefore you can focus 'through' the specimen top to bottom giving a 3-D image of cells	The sections of tissue are very thin. The image of the cells is therefore flat and 2-D

▼ **Table 2** *Advantages and disadvantages*

Advantages: optical microscope	Disadvantages: TEM
Relatively cheap to buy, use, and maintain	Expensive to buy, use, and maintain
Preparing specimens is relatively quick and easy. The preparations are usually free of artefacts	Preparing specimens is relatively lengthy and complicated. The preparations may contain artefacts
Unaffected by magnetic fields	Affected by magnetic fields
Specimens can be observed live or dead	Only dead specimens can be observed because of the vacuum inside the body of the microscope

> **Revision tip**
> Features seen in the cells of tissues prepared for microscopy, but which are not present in living cells, are called artefacts.

> **Summary questions**
>
> 1. Explain how an image forms in the electron microscope. [3]
> 2. Describe how an image is brought into focus in a transmission electron microscope compared with a light microscope. [2]
> 3. What is an artefact? [2]

> **Question and model answer**
>
> **Q** Why is a vacuum necessary inside the body of an electron microscope?
>
> **A** *Electrons are deflected when they strike molecules of the gases which make up air, making it difficult to focus the electron beam. Producing a vacuum within the microscope solves the problem.*

3.3 Microscope measurements and calculations
Specification reference: 3.2.1.3

Graticules and micrometers

- An eyepiece graticule is a glass disc that rests on a rim inside the eyepiece of the optical microscope and is usually marked with a scale 10 mm long, divided into 100 sub-divisions.
- The stage micrometer is a glass slide, placed on the stage of a microscope and viewed through the eyepiece.
- The slide is usually marked with a scale 2 mm long, divided into 200 sub-divisions, each division one hundredth of a mm (0.01 mm or 10 μm).

Worked example
A student calibrating an eyepiece graticule using an objective lens of known magnification saw that the scale lines of the eyepiece graticule and stage micrometer line up at: eyepiece graticule = 30 units, stage micrometer = 20 units.

Given that the distance between each division of the stage micrometer is 10 μm
30 units of the eyepiece graticule = 20 × 10 = 200 μm

Therefore 1 unit of the eyepiece graticule = $\frac{200}{30}$ = 6.7 μm at the magnification of the chosen objective lens. Now that the eyepiece graticule is calibrated the student can measure the actual size of a cell visible in the optical microscope at the magnification of the objective lens used.

Photographs and diagrams

You can also calculate lengths and widths of cells and cell structures from photomicrographs and diagrams, as well as magnification using different combinations of $\frac{I}{A\ M}$

where
I = size of image (measured length)
A = size of object (actual length)
M = magnification

where
$I = A \times M$
$A = \frac{I}{M}$
$M = \frac{I}{A}$

Worked example
Consider a single-celled organism. The scale bar measures 10 mm = 10,000 μm and the size of object (actual length) is given as 5 μm, so M = $\frac{10,000}{5}$ = ×2,000

You can work out the actual size of the specimen by measuring the length x–y

So 2,000 (M) = $\frac{\text{measured } x-y\ (I)}{\text{actual } x-y\ (A)}$

actual x–y (A) = $\frac{\text{measured } x-y\ (I)}{2,000\ (M)}$

measured x–y = 12 mm = 12,000 μm

therefore actual x–y (A) = $\frac{12,000}{2,000}$ = 6 μm. The length of the cell is 6 μm.

Revision tip
Units of measurement
The International System of Units (SI units) is used to measure the size of cells and their structures. The metre (m) is the basic SI unit of length and:

1 m = 1000 mm
1 mm = $\frac{1}{1000}$ m
1 mm = 1000 μm
1 μm = $\frac{1}{1000}$ mm
1 μm = 1000 nm
1 nm = $\frac{1}{1000}$ μm

Revision tip
Any calculation of dimensions of cells and cell structures is possible providing you are given, or can measure, two of the three variables I, A, and M.

Summary questions

1. The illuminated field of view of an optical microscope at × 100 magnification is 2.0 mm. A student observed four whole cells touching each other in line lengthways from one edge of the field of view across its diameter to the opposite edge. Assuming the length of the cells is the same in each case, calculate the actual length of one of the cells. [3]

2. Explain why observing four whole cells improved the accuracy of the student's calculation. [1]

3.4 Eukaryotic cell structure
Specification reference: 3.2.1.1

Structure animal and plant cells seen in the TEM

The term organelles collectively refers to the different structures of eukaryotic cells. Organelles form different compartments within eukaryotic cells. Compartmentalisation allows different activities that might otherwise interfere with one another to occur at the same time within cells.

Mitochondria and chloroplasts

In mitochondria the inner membrane is folded into cristae (sing. crista) which project into the lumen of the matrix of each mitochondrion. In chloroplasts the thylakoid membranes are stacked like pancakes, each stack called a granum (pl. grana).

The folds and stacks increase the surface area of the membranes, maximising the rate of transfer of electrons and protons. Transference of electrons and protons within mitochondria and chloroplasts releases energy that drives the synthesis of ATP.

Endoplasmic reticulum (ER)

There are two types:

Rough ER: so called because of its beaded appearance in the TEM. The 'beads' are ribosomes attached to its membranes. Substances are transported throughout the cell through the network of channels of the rough ER.

Smooth ER: so called because it lacks ribosomes. It consists of flattened sacs that are sites of lipid synthesis.

Cell walls

The wall surrounding the tube-like cells of fungi does not contain cellulose. It consists of glycoproteins and the polysaccharides glycan and chitin: a carbohydrate molecule which also contains atoms of nitrogen as well as atoms of carbon, hydrogen, and oxygen.

The walls surrounding the cells of algae consist of either cellulose, or glycoproteins, or a mixture of glycoproteins and cellulose.

> **Key terms**
>
> **Eukaryotic:** (cells) Having membrane-bound organelles and a distinct nucleus bound by a nuclear membrane (envelope). The cells of plants, animals, fungi, and protists (single-celled organisms, e.g. *Amoeba*) are eukaryotic.
>
> **Nucleus:** Where most of a cell's DNA is found. DNA bound to protein forms chromatin, which becomes visible in the light microscope as chromosomes at the beginning of mitosis. The nucleolus within the nucleus is a region where mRNA is concentrated.
>
> **Golgi apparatus and vesicles:** A stack of sac-like membranes which package different substances (e.g. carbohydrates and proteins forming glycoproteins). Vesicles bud off filled with packaged substances and pass to the cell-surface membrane. Here the substances are secreted (released from the cell).
>
> **Lysosomes:** Contain high levels of lysozymes that destroy old cells and bacteria engulfed in the vacuoles of phagocytes.
>
> **Vacuoles:** Spaces in the cytoplasm of cells. In plant cells vacuoles are permanent, each lined by a membrane called the tonoplast. The vacuole is filled with a solution of sugars and salts which regulate osmosis. In animal cells, vacuoles are temporary and not lined by a tonoplast.

> **Summary questions**
>
> 1. Explain the importance of compartmentalisation in cells. [1]
>
> 2. Describe the role in eukaryotic cells of:
>   a ribosomes
>   b mitochondria
>   c endoplasmic reticulum
>   d Golgi apparatus. [4]
>
> 3. Compare the similarities and differences between mitochondria and chloroplasts. [6]

3.5 Cell specialisation and organisation
Specification reference: 3.2.1.1

Cell differentiation
The cells of an embryo at an early stage all look alike. They are undifferentiated. As the embryo develops, its cells become different from another. They differentiate into particular types of cells. Differentiated cells are specialised, enabling each type to carry out a particular task (its function). For example, producing mucus or conducting nerve impulses.

Tissues
A tissue is a group of cells of the same type. For example epithelia are tissues that cover internal and external surfaces. The appearance of epithelial cells seen with the optical microscope is often a clue to the function of a particular type of epithelium.

- **Protection** of internal organs from damage: for example, the epithelial cells forming the skin are thickened with the protein keratin which helps the cells to resist abrasion.
- **Diffusion** of substances across the surface of the epithelium: for example, the wall of the alveolus is one cell thick. Its cells are flattened, facilitating the diffusion of oxygen and carbon dioxide across its surface.
- **Absorption** of materials: for example, the free surface of the epithelial cells covering the villi of the small intestine are folded into microvilli. The increase in surface area maximises the rate of absorption of digested food.
- **Secretion** of substances onto the surface of the epithelium: for example, goblet cells secrete mucus onto the surface of the epithelium lining the tubes of the trachea and bronchi through which air passes to and from the lungs.

Other different types of cell form other types of tissue. Muscle, liver, and nerve cells are examples.

Organs and organ systems
Different tissues working together make up an organ. Different organs combine to make an organ system.

- The skin is the largest organ in the human body. Other organs include the liver, kidney, heart, stomach, lungs, brain, and ovary.
- Organ systems in the human body include the lymphatic, respiratory, digestive, urinary, reproductive, muscular, skeletal, nervous, endocrine, integumentary (skin, hair, etc.), and circulatory systems.

Plants also have tissues, organs, and organ systems.

> **Summary questions**
>
> 1. Briefly explain the role of genes in the development of the embryo. [3]
>
> 2. Why is the appearance of epithelial cells often a clue to their function? [3]
>
> 3. Briefly explain the difference between a tissue, organ, and organ system. [3]

> **Revision tip**
> More than 200 different types of cell make up the human body. Fewer different types of cell make up the body of a flowering plant.

> **Go further**
>
> Each human being develops from a zygote (a fertilised egg). Development begins when the zygote divides by mitosis. The first few cycles of mitotic division produce a small cluster of cells. These cells give rise to all of the different types of cell of the human body, each type specialised and eventually forming a particular body tissue. The small cluster of cells produced following division of the zygote, and nearly all of the different types of cell that arise from the cluster are genetically identical. Use the information provided and your own knowledge to answer the following questions.
>
> a Explain why the phrase 'the zygote divides by mitosis' is not an accurate statement.
>
> b Which types of human cell are not genetically identical to all of the other types of cell?
>
> c Explain the phrase 'each type (of cell) specialised'. Give three examples of specialised cell and their function.
>
> d How is it possible that most of the types of human cell arising from a zygote are genetically identical to it, but different from it and each other?

3.6 Prokaryotic cells and viruses

Specification reference: 3.2.1.2

Prokaryotic cells

Prokaryotic cells do not have membrane-bound organelles. All bacteria are prokaryotic cells. They are smaller than eukaryotic cells and only just visible in an optical microscope. Their detailed structure is revealed in the TEM.

Many of the features of eukaryotic (plant and animal) cells are not found in prokaryotic (bacteria) cells. Table 1 makes the comparison.

▼ **Table 1** *Plant, animal, and bacterial cells compared*

Feature	Plant	Animal	Bacterium
Cell wall	✔ (cellulose)	✘	✔ (murein)
Nucleus	✔	✔	✘
Plasmids	✘ (except chloroplasts and mitochondria)	✘ (except mitochondria)	✔
Mitochondria	✔	✔	✘
Ribosomes	✔	✔	✔ (but small)
Chloroplasts	✔	✘	✘
Permanent vacuole	✔	✘	✘

> **Revision tip**
> The term acellular literally means *not a cell*. Viruses are not cells.

Viruses

Viruses seem not to need food as a source of energy and they cannot reproduce independently. To do so, viruses must enter a living host cell.

- Attachment proteins at the virus' surface bind to the cell surface membrane of a host cell.
- The virus' genetic material (DNA or RNA) is inserted into the cell.
- The virus' genetic material becomes part of the host cell's genetic material which encodes the synthesis of virus proteins.
- The proteins assemble, forming new virus particles (replication).
- Virus particles escape when the host cell bursts.

They can be crystallised and stored, remaining inactive for years until moistened, which renews their activity.

Most viruses consist of a protein coat, called the capsid, enclosing a strand of nucleic acid (DNA or RNA). A lipid envelope surrounds the capsid of some types of virus (e.g. HIV). Attachment proteins are part of the capsid or lipid envelope (if present).

> **Summary questions**
>
> 1. List the features of prokaryotic cells *not* found in eukaryotic cells. [3]
> 2. How do viruses reproduce? [4]
> 3. Using a bacterium and virus as examples, distinguish between living and non-living things. [5]

3.7 Mitosis
Specification reference: 3.2.1.1

Mitosis
During mitosis replicated DNA (in the form of chromatids) of the parent cell appears as distinct chromosomes under the optical microscope.

- The movements of the chromatids during mitosis in a plant cell are identical to those in an animal call, except that spindle formation takes place in the *absence* of centrioles (spindle formation, therefore, does not depend on centrioles).
- A cell inherits two sets of chromosomes, which is why the parent cell and its daughter cells are said to be diploid: one set from the male parent, the other set from the female parent.
- Mitosis increases cell numbers in an organism. The growth and repair of tissues depends on mitosis.

DNA replication
Remember that DNA replication occurs during the S phase of interphase followed by mitosis and cytokinesis.

- As a result the two daughter cells each inherit an exact copy of the DNA of the parent cell.
- As a result daughter cells and parent cells are genetically identical.

This is why growth and the repair of tissues depends on mitosis: old skin cells divide forming identical new skin cells; old gut cells divide forming identical new gut cells and so on.

A group of genetically identical DNA molecules / cells / organisms is called a clone. Asexually reproduced organisms from a single parent form a clone because asexual reproduction depends on mitosis.

Cytokinesis
Cytokinesis follows division of the nucleus at the end of telophase.

- In animal cells filaments of actin attached to the inner surface of the plasma membrane contract and pull the membrane inwards. A **division furrow** forms which deepens, eventually splitting the parent cell into two daughter cells.
- In plant cells a **cell plate** forms which extends outwards until it meets the plasma membrane. A new cell wall forms, splitting the parent cell into two daughter cells.

> **Revision tip**
> Colchicine is a chemical extracted from autumn crocus corms. Metaphase can be seen more clearly if colchicine is added to a culture of cells undergoing mitosis. Spindle formation is inhibited preventing the separation of chromatids during anaphase.

> **Revision tip**
> Feulgen's solution stains DNA dark purple. Staining mitotic cells with Feulgen's solution makes it easier to visualise their chromosomes through the optical microscope.

> **Synoptic link**
> The replication of DNA was covered in Topic 2.2, DNA replication.

> **Summary questions**
>
> 1 Why are the daughter cells of a parent cell dividing by mitosis genetically identical to one another? [3]
>
> 2 Explain why cytokinesis in a plant cell is not the result of the formation of a division furrow but by a cell plate. [2]

3.8 The cell cycle
Specification reference: 3.2.2

The cell cycle

New cells are formed from existing cells.

- The cells that give rise to new cells are called parent cells.
- The new cells are the daughter cells formed when the parent cell divides.

The cell cycle describes the process. There are three stages: interphase, mitosis, and cytokinesis. Mitosis and cytokinesis are described in Topic 3.7, Mitosis. In this topic the focus is on interphase.

Interphase

The genetic material in the cell's nucleus appears under the optical microscope as diffuse chromatin. The different phases of interphase are called G_1, S, and G_2.

When mature (fully differentiated), some types of cell do not divide (quiescence). Such cells enter the G_0 phase of the cell cycle. Permanent quiescence represents a checkpoint where cell division is inhibited in response to damaged DNA. As a result harmful mutations are not inherited by daughter cells. Temporary quiescence may be due to the lack of nutrients needed for cell division. When more nutrients become available, then the cells re-enter G_1.

Broadly speaking regulation is either:

- inhibition: cells are prevented from proceeding to the next stage of the cell cycle

or

- stimulation: cells are able to proceed to the next stage of the life cycle.

> **Revision tip**
> Cell cycle checkpoints control the progression of cells from one stage (or phase) of the cell cycle to the next.

> **Revision tip**
> The mutated gene that results in a cancer is called an oncogene. There are more than 200 different types of cancer.

What causes cancer?

A cancer develops when the events of the cell cycle run out of control. Loss of control occurs when the genes regulating what happens at cell cycle checkpoints mutate.

Chemical factors in the environment and ionising radiation are potential carcinogens (cause cancer).

Chemicals

Many of the chemicals in the tar of tobacco smoke are carcinogens.

- Tobacco carcinogens cause the mutation of tumour suppressor genes which normally inhibit cell division. The carcinogens inactivate the genes.
- Normally the genes 'switch on' when their task is done. If tobacco carcinogens cause the genes to mutate to oncogenes which do not 'switch on', cells divide uncontrollably and cancers develop.

Ionising radiation

γ (gamma)-rays and X-rays are ionising radiations. They damage DNA causing mutations which may lead to the development of a cancer.

Treating cancer

If detected at an early stage of development, many cancers can be successfully treated by:

- **Surgery:** the cancer is removed.
- **Chemotherapy:** drugs which block different stages of the cell cycle, for example different drugs target:
  - DNA replication preventing it from happening
  - metaphase during mitosis by interfering with spindle formation.
- **Radiotherapy:** the cancer is bombarded and destroyed by a stream of particles emitted by a radioactive source.
- **Immunotherapy:** some of the proteins which are part of the cell-surface membrane of cancer cells are different (abnormal) from the cell-surface proteins of healthy cells in the same person. These abnormal proteins are antigens. Immunotherapy makes it easier for the person's immune system to detect the antigens. As a result the immune system mounts an immune response which destroys the cancerous cells.

Other new treatments include:

- **Vaccines** against cancer-causing viruses (human papilloma viruses cause cervical cancer).
- **Gene therapy** replaces the faulty genes which cause cancer with healthy genes.
- **Monoclonal antibodies** target specific abnormal cell-surface proteins of cancer cells. When combined with an anti-cancer drug, the monoclonal antibodies deliver the drug to the cancer cells without affecting healthy cells.

Left untreated, cancerous cells can break away from the original (primary) tumour, spread (a process called metastasis), and set up secondary growths elsewhere in the body, endangering the person's life.

> **Summary questions**
>
> 1 Cancer is often described as the result of the cell cycle 'run out of control'. Explain the statement. [3]
>
> 2 When may a mutagen not be a carcinogen? [1]
>
> 3 Explain why cancer treatments based on immunological methods are more likely to minimise damage to healthy tissue. [5]

Chapter 3 Practice questions

1 The figure represents a mitochondrion seen in a transmission electron microscope.

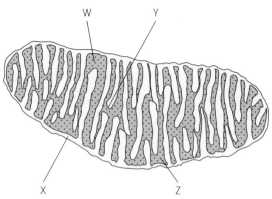

   a Name the features labelled X, Y, and Z. *(3 marks)*

   b Name the feature labelled W. State its function. *(2 marks)*

   c There are a number of structures identified by label Y visible in the photograph. Some of the structures are attached to the inner mitochondrial membrane and others are not. Explain why. *(2 marks)*

2 The diagram represents a type of bacterium that causes food poisoning.

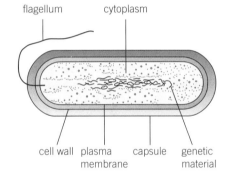

   a Give one similarity and identify three differences between the structure of the bacterial cell and the structure of a palisade cell in a leaf. *(4 marks)*

   b How is the genetic material in bacterial cells organised? *(2 marks)*

   c The cell wall of the bacterium consists of protein, lipid, and a polysaccharide called murein. Name two polysaccharides not found in the cell wall of bacteria but found in the cell wall of the cells of fungi. *(2 marks)*

   d State one unusual feature of a molecule of murein. *(1 mark)*

3 During mitosis, a parent cell gives rise to two daughter cells. The cell cycle describes the process.

   a The mass of DNA of a cell entering interphase is 18 arbitrary units. The mass of the same cell entering mitosis is 36 arbitrary units. Explain how the mass of the cell has doubled. *(2 marks)*

   b Explain why 'cells divide by mitosis' is not an accurate statement. *(3 marks)*

   c The figure represents cells during mitosis. Figure **a** shows cells in metaphase; Figure **b** shows cells in anaphase.

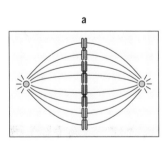
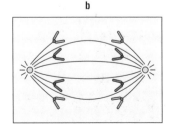

      i Using Figures **a** and **b**, explain how you can identify each phase from the appearance of the chromosome. *(3 marks)*

   The figures show the chromosomes attached to spindle fibres.

      ii In animal cells, name the structures which give rise to the spindle fibres. *(1 mark)*

      iii How does the origin of spindle fibres in plant cells differ from animal cells? *(1 mark)*

   d Checkpoints regulate the progression of cells from one stage of the cell cycle to the next. Different genes express different enzymes which help to regulate what happens at the checkpoints. Broadly speaking, regulation either inhibits or stimulates the progression of cells.

      i Explain how mutation of the genes regulating checkpoints in the cell cycle might lead to the development of a cancer. *(5 marks)*

      ii Summarise how increasing understanding of the cell cycle is helping in the development of new drug treatments for cancer. *(5 marks)*

4.1 Structure of the cell-surface membrane
Specification reference: 3.2.3

Structure and functions of the cell-surface membrane

Most of the organelles of the eukaryotic cell are formed from membranes. The eukaryotic cell itself is surrounded by the cell-surface membrane. A cell-surface membrane also surrounds the prokaryotic cell. Cell membranes:

- are partially permeable barriers which regulate the movement of solutes within the cell and between the cell and its environment
- separate the cell from its surroundings, allowing control of its internal environment
- in eukaryotic cells, compartmentalise the contents of organelles from the rest of the cell
- may be folded increasing their surface area. This means the rate of movement of molecules / ions across membranes (e.g. diffusion) is maximised.

Phospholipids

Phospholipids are an important part of the cell-surface membrane. They have hydrophilic heads (phosphate groups) and hydrophobic tails (hydrocarbon chains). Cell contents are aqueous and cells are bathed in water. As a result, in water phospholipid molecules spontaneously form a stable two-layer framework called a phospholipid bilayer.

Hydrocarbon tails

The hydrocarbon tails of many of the phospholipid molecules are 'kinked' because of the presence of double bonds. As a result phospholipids are loosely packed together, membranes are fluid, and proteins and other material can move sideways.

Cholesterol

Cholesterol is a lipid. Its molecules are attracted to the hydrocarbon tails of the phospholipid molecules of membranes and insert between them. As a result the membrane is slightly immobilised, firming it and restricting the movement of very water-soluble substances across it. Without cholesterol cell membranes would be too fluid.

Phospholipid bilayer

Different proteins are embedded in the phospholipid bilayer. Some proteins are integral proteins and extend through the membrane; peripheral proteins are localised on one of the sides of the membrane.

Functions of the cell-surface membrane

The different proteins perform many of the functions of the cell-surface membrane. Functions include:

- action as enzymes – for example, enzymes on the surfaces of cells lining the intestine catalyse reactions which digest food
- transport of substances across the membrane
- maintenance of cell shape
- binding of messenger molecules. The proteins to which messenger molecules bind are called receptors. The messenger molecules trigger particular activities in the cell.

Revision tip
Compartmentalisation allows many different activities to occur simultaneously and independently within a cell.

Revision tip
Sugars bond to proteins (as glycoproteins) and lipids (as glycolipids) embedded in the outside surface of the plasma membrane. Glycolipids and glycoproteins are receptors for chemicals that are signals between cells.

Summary questions

1 Outline the roles of membranes in eukaryotic cells. [3]

2 Explain why the membranes of a cell are described as a fluid mosaic. [2]

4.2 Diffusion

Specification reference: 3.2.3

> **Revision tip**
> Cell membranes allow some molecules and ions to pass *freely* across but *restrict* the passage of others. We say that the membranes are partially permeable. Restriction occurs because the hydrophobic region of the phospholipid bilayer is more viscous than water.
>
> Rates of diffusion depend on the polarity and size of particles of substance. Non-polar lipid-soluble molecules diffuse more rapidly across membranes than large polar molecules (e.g. globular proteins).

Diffusion

Diffusion is the net movement of a substance through a gas or solution from a region where the substance is in high concentration to a region where it is in low concentration. Diffusion continues until the concentration of the substance is the same throughout the gas or solution.

Different factors affect the rate of diffusion:

- The concentration gradient of a substance is the difference in concentration between the regions through which the substance is diffusing (diffusion pathway), divided by the distance.
- The greater the difference between regions of high and low concentration of a substance, the greater the concentration gradient. As a result the rate of diffusion is maximised.
- The rate of diffusion of a substance decreases in proportion to the square of the distance over which diffusion is taking place (rate $\propto \frac{1}{\text{distance}^2}$).
   - As a result diffusion is only effective as a mechanism for the transport of substances over very short distances.
   - As a result membranes are thin (diffusion pathway is short).
   - As a result the size of cells is limited because diffusion is an important mechanism for the transport of substances across membranes within cells.
- The greater the surface area of a membrane, the greater is the rate of diffusion of substance through it.

Facilitated diffusion

Many types of hydrophilic molecules (and ions) diffuse into and out of cells through pores formed by different carrier proteins which cross from one side of the plasma membrane to the other (integral proteins). The process is called facilitated diffusion.

The pores of the different carrier proteins are filled with water, forming a hydrophilic channel through the hydrophobic region of the phospholipid bilayer. So hydrophilic molecules (and ions) can pass through the channels and therefore through membranes more easily.

Pores are specific to the substances that pass through them. The rate at which a substance diffuses through the pores of a carrier protein depends on the steepness of its concentration gradient across the membrane, the type of carrier protein forming the pore, the number of pores in the membrane, and whether the pores are open or not.

> **+ Go further**
> Explain how changes in temperature affect the function of cell membranes.

Summary questions	
1 Explain why the process of diffusion limits the size of cells.	[2]
2 Explain how carrier proteins in cell membranes are each specific to the substance it passes across a membrane.	[2]
3 Define the term 'concentration gradient'.	[2]

4.3 Osmosis

Specification reference: 3.2.3

What is osmosis?

Osmosis is the net movement of water molecules through a partially permeable membrane from a region where they are in a higher concentration to a region where they are in a lower concentration. The term refers to the diffusion of water molecules and is *only* used in this context.

Water potential

Water molecules in motion striking a membrane exert pressure called the water potential. The higher the concentration of water molecules, the greater is the kinetic energy of the system and the greater is the water potential. Water potential is measured in kilopascals (kPa).

Thinking about water potential

The water potential of pure water at atmospheric pressure is zero (0). The addition of a solute such as glucose lowers the water potential because there is a lower concentration of water molecules per unit volume of the solution. As a result the water potential of the solution has a negative value. The more solute is dissolved, the greater is the negative value.

Using symbols

ψ (sometimes written as ψ_w or ψ_o) = the water potential of a solution.

ψ_s = the solute potential which is a measure of the extent to which solute concentration lowers the water potential of a solution.

ψ_p = the pressure potential which is a measure of the pressure that develops inside a cell as a result of the inflow of water.

> **Revision tip**
>
> **Osmosis in terms of water potential**
>
> Osmosis is the movement of water down a water potential gradient across a partially permeable membrane from a solution of a higher (less negative) water potential to a solution of a lower (more negative) water potential.

> **Revision tip**
>
> For animal cells where absence of a cell wall means that the pressure potential must be negligible, otherwise the cells would burst:
>
> $\psi = \psi_s$ ψ_p is zero
>
> For plant cells where the cell wall prevents cells from bursting because of the build up of internal hydrostatic pressure:
>
> $\psi = \psi_s + \psi_p$ ψ_s is a negative value since the ψ for pure water is zero
>
> ψ_p is a positive value

Worked example

The table gives the values for ψ_s and ψ_p of adjacent plant cells A and B. Calculate the water potential for each cell and predict the direction of water movement between the cells.

	Cell A	Cell B
	$\psi_s = -1800$ kPa	$\psi_s = -1250$ kPa
	$\psi_p = 650$ kPa	$\psi_p = 600$ kPa

For cell A

$\psi = -1800 + 650$

$= -1150$ kPa

For cell B

$\psi = -1250 + 600$

$= -650$ kPa

The water potential of cell B is greater (less negative) than cell A, so water will move down the water potential gradient from cell B to cell A.

Summary questions

1. Why might osmosis be described as a special case of diffusion? [2]

2. Briefly describe osmosis in terms of water potential. [2]

3. Explain why water potential is more negative the more solute there is in solution. [3]

4.4 Active transport

Specification reference: 3.2.3

Active transport

Sometimes molecules or ions move across membranes from where they are in lower concentration to where they are in higher concentration. In other words they move in the reverse direction to diffusion. The process is called active transport and allows cells to build up stores of soluble substances that would otherwise be spread out by diffusion. The storage of glucose by liver cells is an example.

The process is active in the sense that more energy is required to move the molecules or ions against their concentration gradient than down it. This is the energy released by the hydrolysis of ATP, which is produced during cellular respiration.

Diffusion, facilitated diffusion, and osmosis are passive processes in the sense that energy provided by the hydrolysis of ATP is not required. The kinetic energy provided as the result of the kinetic motion of molecules is sufficient to move molecules or ions down their concentration gradient.

Although active transport requires ATP and facilitated diffusion does not, the movement of molecules and ions by both processes is achieved by carrier proteins. Each type of carrier protein is specific for a particular molecule or ion.

The sodium–potassium pump is an example of a carrier protein which actively transports sodium (Na^+) ions and potassium (K^+) ions across the plasma membrane of cells.

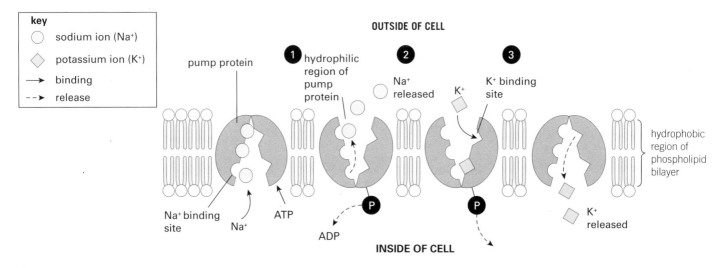

▲ **Figure 1** *The sodium–potassium pump. Carrier proteins which transport substances in opposite directions are called antiports*

- The concentration of sodium ions (Na^+) is higher outside the cell than inside.
- The concentration of potassium ions (K^+) is higher inside the cell than outside.
- Three Na^+ and one molecule of ATP bind to the pump protein.
- The ATP is hydrolysed and ADP is released. The phosphate group remains bound to the pump protein. As a result the shape of the pump protein changes and the three Na^+ pass from the cell against the concentration gradient of Na^+ and are released.
- Two K^+ bind to the pump protein.
- The phosphate group bound to the pump protein is released. As a result the structure of the pump protein changes back to its original shape and the two K^+ pass into the cell against the concentration gradient for K^+ and are released.

Transport across cell membranes

Endocytosis and exocytosis

Large molecules such as proteins cross membranes by a process called cytosis.

Energy released by the hydrolysis of ATP to ADP is required, and the process is made possible by the fluidity of the membrane. Exocytosis – a vesicle containing molecules of substance fuses with the inside of the cell surface membrane and the molecules are secreted (removed) from the cell. Endocytosis – the cell surface membrane binds to molecules: a vesicle sac forms. The sac enters the cells.

A vesicle (sac) forms around the molecules, and the sac enters the cell. Phagocytosis (uptake of solids) and pinocytosis (uptake of fluids) are examples of endocytosis.

▼ **Table 1** *Summary of diffusion, active transport, and endocytosis*

Process	Proteins	ATP/energy	Concentration gradient	For example
Diffusion	May be pore proteins	No	Down	Water, Na^+
Facilitated diffusion	Yes	No	Down	Glucose, amino acids in gut
Active transport	Yes	Yes	Up or down (usually up)	Na^+ out of neurones
Endocytosis	Sometimes, to recognise the molecule	Yes	Up or down (usually down)	Milk proteins across baby's gut wall

Summary questions

1. Explain the differences between the processes of diffusion, facilitated diffusion, and active transport. [2]

2. Identify and explain the role of the sources of energy required for active transport. [2]

4.5 Co-transport and absorption of glucose in the ileum

Specification reference: 3.2.3

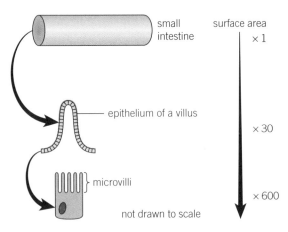

▲ Figure 1 Increasing surface area

Surface area

The inner surface of the cells of wall of the ileum (small intestine) are folded. The folds are called villi and increase the surface area available to absorb the products of digestion. Figure 1 shows how these different features affect the orders of increase in surface area, compared with the internal surface area of a cylinder.

The surface of the cells of the epithelium lining the villi is thrown into tiny folds, visible only in the electron microscope. The folds are called microvilli and form what is often called the brush border because they look a little like the bristles of a brush.

Absorption

Absorption of the products of carbohydrate digestion begins in the duodenum, but most occurs across the wall of the ileum. It starts at the brush border of microvilli at the surface of each villus.

Molecules of sugar pass:

- through the cell-surface membrane of each epithelial cell
- across each cell
- through the cell-surface membrane on the opposite side of each cell
- through the wall of a capillary blood vessel or the wall of a lacteal vessel.

These vessels are parts of a network of vessels within each villus. The capillaries run into the hepatic portal vein; the lacteal vessels run into the lymphatic system.

Passage of the molecules across the cell-surface membrane of the cells is by diffusion, facilitated diffusion, and active transport. Figure 2 shows the process.

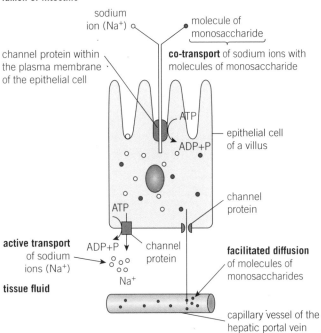

▲ Figure 2 Absorption in the intestine

- **Glucose** and other sugars are some of the products of digestion which pass into the hepatic portal vein to be transported in the bloodstream to the liver.
- **Carrier proteins** in the cell-surface membrane couple the transport of monosaccharides with sodium ions (Na^+) from the lumen of the intestine into the epithelial cells of the villi. The process is called co-transport because the movement of monosaccharides and sodium occurs together. The process is a form of active transport requiring ATP as a source of energy.
- **Monosaccharide** molecules transfer from the epithelial cells into capillary vessels of the hepatic portal vein by facilitated diffusion.
- **Sodium ions** (Na^+) are pumped from the base of epithelial cells into the surrounding tissue fluid by active transport. ATP is required as a source of energy.

Summary questions

1 Summarise the features of the small intestine which increase its internal surface area compared with the internal surface area of a cylinder. [2]

2 Briefly explain how sugar molecules are absorbed by a cell lining the wall of the ileum (small intestine). [2]

Chapter 4 Practice questions

1 The diagram shows a molecule of phospholipid:

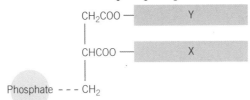

   a Using the information in the figure and your own knowledge, explain how phospholipid molecules in water spontaneously form a bilayer.
      (2 marks)

Compare the hydrocarbon chains of the fatty acids X and Y in the figure.

   b Identify the difference between X and Y and explain the significance in terms of the structure of cell membranes. *(2 marks)*

   c The lipid cholesterol is also part of the structure of cell membranes. It helps to maintain the flexibility and shape of the membranes. Explain how. *(2 marks)*

2 In times of war, opponents try to disrupt the other side's food supplies. For example, the Romans scattered salt in the wheat fields of their opponents, destroying the plants. Repeating the experiment today and sampling the tissues of plants treated and untreated with salt, the cells of sampled tissues seen under the optical microscope would appear like this:

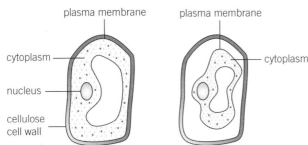

   a Describe the appearance of the cytoplasm of each cell. *(2 marks)*

   b Explain in terms of water potential the differences in the appearance of cells A and B. *(3 marks)*

Let ψ symbolise the water potential of a cell and ψ_s its solute potential.

   c In animal cells explain why

      $\psi = \psi_s$ *(2 marks)*

Let ψ_p symbolise the pressure potential of a cell.

   d In plant cells $\psi = \psi_s + \psi_p$

      In terms of the equation and the properties of cellulose, explain how ψ_p enables a plant cell to become turgid. *(3 marks)*

3 Molecules of different substances are transported across membranes by facilitated diffusion and active transport. Both processes require carrier proteins. Each substance is specific to its particular carrier protein.

   a Explain why active transport requires ATP and facilitated diffusion does not. *(1 mark)*

   b In terms of the tertiary structure of proteins, explain the basis of specificity of a substance transported by its particular carrier protein. *(2 marks)*

   c Explain why the molecules of a substance similar in shape to the molecules of the substance normally transported by its specific protein carrier inhibit the transport of that substance. *(3 marks)*

5.1 Defence mechanisms

Specification reference: 3.2.4

> **Synoptic link**
> Topic 5.2, Phagocytosis, deals with the first line, non-specific defences of the innate immune system. Topics 5.3–5.5 deal with the defences of the adaptive immune system.

Defence mechanisms

The body's defences against disease consist of the:

- **innate immune system:** defence is immediate and non-specific (the same for all threats to the health of the individual)
- **adaptive** (or **acquired**) **immune system:** defence is long term and specific (defence is individual to each health threat).

Self and non-self

Many different types of protein with different functions are components of the phospholipid bilayer forming cell membranes.

How does the human body distinguish between its own cells (self) and cells which are not, abnormal self body cells, pathogens, and toxins (non-self)?

Some of the different proteins of cell membranes are marker molecules enabling cells of a body to recognise one another as self. Human marker molecules are encoded by genes called the major histocompatibility complex (MHC).

- Some self-marker molecules are class 1 MHC antigens. They enable:
  - cells of the same body to recognise one another
  - the lymphocytes and phagocytes to recognise each other and not mount an immune response
  - T cells to mount an immune response against abnormal cells.

Other self-marker molecules are class 2 MHC antigens. B cells and phagocytes are **antigen presenting cells**. They present (highlight) non-self antigens to **T helper (T_H)** cells. These stimulate **T cytotoxic (T_C)** cells to destroy foreign antigens.

Recognising self

Self-tolerance enables the body to distinguish between self and non-self antigens.

- **Central tolerance:** T cells and B cells that develop surface receptors which recognise self-antigens are destroyed or inactivated before they can mount an immune response against self-antigens. Their removal through apoptosis (cell death) is called clonal deletion.
- **Peripheral tolerance:** Immature T cells and mutated B cells can recognise self-antigens as well. Clonal deletion removes the cells and therefore the threat of the cells mounting an immune response against self-antigens.

If clonal deletion/anergy do not work properly then autoimmune disease may result. The body's immune system destroys its own tissues. Type 1 diabetes is an example.

Non-self and rejection

MHC genes and the proteins they encode are different between individuals (except identical twins). As a result the marker molecules of cells from one person will be identified as non-self (foreign) by the cells of another person (except an identical twin). This is why transplanted tissues and organs are at risk of a response from T cells of the immune system and rejection by the transplant recipient's body.

> **Revision tip**
> Sometimes, T cells carrying receptors for self-antigens are not deleted. Instead they develop a lack of responsiveness to the antigens. The process is called clonal anergy.

> **Summary questions**
> 1. Distinguish between the innate immune system and the adaptive immune system. [2]
> 2. What is an immune response? [1]
> 3. Explain the difference between *central tolerance* and *peripheral tolerance*. [4]

5.2 Phagocytosis
Specification reference: 3.2.4

Phagocytosis

The term phagocytosis derives from ancient Greek and literally means 'cell eating' (engulfing cells). It is one of the body's first line defences against disease. Other first line defences are physical (P) and chemical (C):

- the skin is a barrier to water and pathogens (organisms that cause disease) (P)
- sebaceous glands in the skin produce sebum, an oily liquid that destroys pathogens (C).

Natural openings in the skin are also defended. For example, tear glands of each eye produce lysozyme, the enzyme that catalyses reactions that destroy pathogens (C). Damage to the skin is quickly sealed by blood clotting; a scab forms a barrier over the wound (P).

Flexibility of the cell-surface membrane enables phagocytes to engulf particles including bacteria and viruses that cause disease (pathogens). See Figure 1.

> **Revision tip**
> Cell membranes are flexible. They are often described as a fluid mosaic. Phagocytosis is one of the defence mechanisms of the innate immune system. Phagocytes are a type of white blood cell.

> **Synoptic link**
> Look back at Topic 3.4, Eukaryotic cell structure, to remind yourself of the role of lysosomes in the cell.

1 The phagocyte is attracted to the pathogen by chemical products of the pathogen. It moves towards the pathogen along a concentration gradient

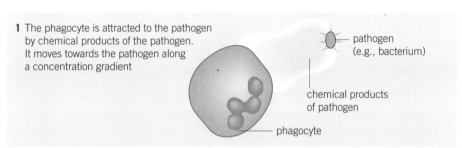

2 The phagocyte has several receptors on its cell-surface membrane that attach to chemicals on the surface of the pathogen

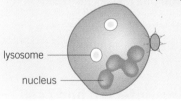

3 Lysosomes within the phagocyte migrate towards the phagosome formed by engulfing the bacterium

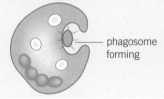

4 The lysosomes release their lysozymes into the phagosome, where they hydrolyse the bacterium

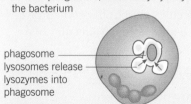

5 The hydrolysis products of the bacterium are absorbed by the phagocyte

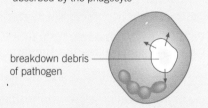

▲ Figure 1 Summary of phagocytosis

Chemicals released by pathogens attract phagocytes to them because non-specific cell-surface receptors (proteins) of phagocytes detect the antigens on pathogens. The chemicals form a concentration gradient down which phagocytes move to where the pathogens are. The process is an example of chemotaxis.

Receptors bind with antigens causing the receptors to cluster, which triggers phagocytosis.

> **Revision tip**
> Proteins (e.g. IgG, a type of antibody) that bind to the surface antigens of pathogens enable phagocytes to recognise pathogens more easily. These proteins are called opsonins. Opsonisation speeds up phagocytosis, making it more effective.

> **Summary questions**
>
> 1 Describe the role of opsonins. [1]
>
> 2 Explain how phagocytes are attracted to pathogens. [2]
>
> 3 Describe the interaction between a phagosome and lysosomes. [2]

5.3 T lymphocytes and cell-mediated immunity

Specification reference: 3.2.4

T cells originate in the bone marrow from multipotent (adult) stem cells. They mature in the thymus (hence the name T cells). T cells only respond to foreign antigens on the surfaces of other cells (antigen presenting cells).

Their responses are components of the cell-mediated (cellular) response of the immune system to antigens. T cells do not respond to antigens in body fluids.

T_H cells

HIV infects T_H cells and shows how important T_H cells are to good health. The number of T_H cells decreases with an increasing load of HIV with the result that the symptoms of AIDS (acquired immune deficiency syndrome) appear. The symptoms are different diseases which are normally rare. They occur in HIV positive people because destruction of T_H cells by HIV means that the immune system destroys the causes of the diseases less effectively. Treatment is the prescription of a cocktail of drugs (combination therapy) that reduces the infected person's HIV load.

T_H cells:

- activate and promote proliferation of T_C cells
- produce cytokines (e.g. interferon)
- produce opsonins that cover antigens making it easer for phagocytes to recognise and target the antigens
- promote proliferation of B cells
- increase production of antibodies by B cells.

The effects of T_H cells on other T cells, B cells, and phagocytes ensure a healthy immune system.

T_C cells are also known as T-killer cells and destroy virus-infected cells and cancerous cells.

Events when a T_C cell responds to a virus-infected body cell

A T_C cell binds to virus antigen on the surface of the virus-infected cell (target cell). Perforin and granzymes (A and B) are proteins produced by the T_C cell and secreted by exocytosis. Perforin molecules bind to the cell-surface membrane of the target cell and perforate the membrane. Molecules of granzymes A and B pass into the target cell:

- granzyme A catalyses reactions that poison the target cell
- granzyme B catalyses reactions that destroy the target cell.

Before its destruction, the target cell produces and releases a protein called **interferon**.

Interferon inhibits virus replication. It also stimulates the anti-viral defences of cells nearby. The virus-infected target cell is destroyed releasing virus particles.

T_H cells stimulate proliferation of B cells and their production of antibodies. The viruses released by the destruction of the target cell are antigens which are targeted by the antibodies.

> **Revision tip**
> T cells and B cells are the two main categories of lymphocyte. T cells do *not* produce antibodies; B cells do. An antigen is any substance that enters the body and stimulates an immune response.

> **Synoptic link**
> You can read more about antigen presenting cells in Topic 5.1, Defence mechanisms.

> **Revision tip**
> T helper (T_H) and T cytotoxic (T_C) cells are examples of types of T lymphocyte.

> **Summary questions**
>
> 1. How do T cells detect antigens? [2]
> 2. Describe the functions of T helper cells. [3]
> 3. How do the actions of molecules of perforin and granzymes destroy virus-infected cells? [3]

5.4 B lymphocytes and humoral immunity
Specification reference: 3.2.4

B cell response to antigens and phagocytosis

B cells divide and produce clones of cells (clonal expansion) in response to the detection of antigens. Most of the cells are plasma cells. They produce antibodies which combine with the antigens. The combinations are called immune complexes.

If immune complexes form between antibodies and the antigenic proteins on the cell-surface membrane of bacterial cells, agglutination (clumping together of cells) occurs. Agglutination helps phagocytes to destroy bacteria (phagocytosis).

Immunological memory

On first encounter with an antigen, clonal expansion takes a few days to produce B plasma cells and T plasma cells (the primary immune response) against the infection. Plasma cells are short lived. But on encountering the same antigens again, the response is much quicker (the secondary immune response). This is because the first clonal expansion also produces memory cells.

B-memory cells rapidly divide to produce plasma cells which then produce antibodies. T-memory cells rapidly divide to produce plasma cells which then take part in cell-mediated immunity responses. These enhanced responses are evidence of immunological memory.

Memory cells are specific for a particular antigen. This is why we rarely catch mumps or chicken pox more than once in a lifetime: the rapid response as a result of immunological memory destroys the viruses which cause these diseases before they make us ill. Memory cells produced because of infection by a pathogen provide long-term immunity.

Antigenic variability in the influenza virus

The surface proteins of influenza viruses are antigens against which an infected person produces antibodies. But people may catch influenza more than once during their lifetime. So why are the antibodies ineffective against influenza virus when it next invades the body? After all, the production of memory cells means that we rarely catch diseases like chicken pox and measles more than once.

Unfortunately, frequent mutation means that influenza virus antigens often change shape.

- Minor changes are called antigenic drift. They produce new strains of virus.
- Major changes are called antigenic shift and result in new types of virus.

Antigenic drift and antigenic shift means that antibodies produced against a particular type of influenza virus do not protect the person from infection by new variants of the virus. The antigens on each new variant influenza virus are different from those on previous variants, meaning that vaccines against previous variants are no longer effective against new ones.

> **Revision tip**
> B cells are components of the humoral response of the immune system to antigens. The term humours is the old fashioned word used to refer to body fluids.

> **Common misconception**
> B cells originate in the bone marrow and mature in the spleen and lymph nodes. Unlike T cells, they do *not* mature in the thymus.

> **Summary questions**
> 1 Distinguish between the responses of B cells and T cells. [2]
> 2 With reference to B cells, what is clonal expansion? [2]
> 3 Describe the significance of memory cells to continuing good health. [3]
> 4 Distinguish between antigenic drift and antigenic shift in the influenza virus. [2]

5.5 Antibodies

Specification reference: 3.2.4

> **Revision tip**
> B plasma cells produce antibodies in response to foreign antigens detected in body fluids.

> **Synoptic link**
> For more information on the ELISA test see Topic 5.7, The human immunodeficiency virus (HIV).

> **Revision tip**
> Complementarity between antigen and antigen binding site is why antibodies are specific to each type of antigen.

> **Revision tip**
> Cancer cells produce antigens different from those produced by healthy cells. Monoclonal antibodies attached to anti-cancer drugs target cancer cells without affecting healthy cells.

> **Summary questions**
>
> 1. **a** Distinguish between the constant region and variable region of an antibody molecule. [2]
>   **b** Explain the significance of the variable region for antibody specificity. [2]
> 2. What is a monoclonal antibody? Explain why monoclonal antibodies have the potential to reduce harmful side effects to cancer patients. [3]
> 3. How do antibodies destroy antigens? [2]

Antibodies

Antibodies are glycoproteins. Each molecule is typically represented as y-shaped with four polypeptide chains making up the molecule's basic structure: 2 long *heavy* chains and 2 short *light* chains.

The bulk of the polypeptide chains is very similar in all antibodies. This part of the molecule is called the constant region. However a small region of the polypeptide chains at one end of an antibody molecule is very variable. This part of the molecule is called the variable region. The region forms the antigen-binding site.

Specificity

Each of us encounters millions of different types of foreign antigen molecule in a lifetime.

- The quaternary structure of the antigen binding site of antibodies varies. As a result there are millions of variant binding sites. Each variant binding site has a unique shape.
- Each type of foreign antigen molecule also has a unique shape.
- When a particular shape of antigen binding site encounters a molecule of foreign antigen with a shape which matches (is complementary to) that of the antigen binding site then the antigen binds with the antigen binding site forming an immune (antigen–antibody) complex.

How do antibodies destroy antigens?

The effect of antibodies is indirect. Instead they make the destruction of antibodies more likely.

- **Agglutination** makes it easier for phagocytes to locate bacteria and destroy them.
- **Precipitation** of antigen facilitates phagocytosis.
- **Markers**: immune complexes are signals that stimulate phagocytosis.
- **Neutralisation** of toxins.
- **Lysis** enzymes bind to antibodies and catalyse reactions which break down bacteria when bound to the antibodies.

Monoclonal antibodies

Separating antibodies into pure samples of particular types of antibody is difficult. Growing B cells in culture is not possible. However, by fusing B cells that produce a particular antibody with a type of rapidly dividing lymphocyte cancer cell called a **myeloma**, the fused cells grow in culture and produce only the required antibody. The fused cells are hybridomas (two different cells fused together). Pure samples of antibodies made in this way are monoclonal antibodies. Monoclonal antibodies have a wide range of uses, including the treatment of some cancers.

Monoclonal antibodies are also put to other uses.

- Monoclonal antibodies bind with poisons, inactivating them.
- The success of transplants depends on matching the tissue of donor and patient as nearly as possible. Monoclonal antibodies, produced against the donor's and patient's tissue antigens, make tissue matching more accurate and therefore tissue rejection less likely.

Developing different types of monoclonal antibody for medical uses is expensive. However with technical and commercial success, treatments based on monoclonal antibody technologies should become less expensive.

5.6 Vaccination

Specification reference: 3.2.4

Vaccines

Natural and artificial immunity

Natural immunity results from the body's response to an antigen. Natural immunity may also be passed from mother to baby as antibodies in her breast milk. Artificial immunity is brought about by vaccinations. A vaccine is a preparation of dead or attenuated (weakened) pathogens or harmless components of pathogens which are foreign antigens that stimulate an immune response in a person receiving the vaccine. It can be given by mouth or by injection.

Active immunity

Following vaccination, the person is protected from the effects of the active form of the pathogen should it infect the body. We say that the person is immune to the pathogen. The immunity is active because the person's immune system has been stimulated by the harmless antigens to produce memory cells. These cells are the basis of the person's immunity. Active immunity is long-lasting.

Passive immunity

Passive immunity comes from the injection of antibodies produced by another animal (e.g. horse). Protection against the effects of a particular pathogen is immediate. However, protection is short term because the immune system of the person vaccinated with antibodies from another animal detects and recognises the antibodies as foreign antigens. The person's immune system mounts an immune response against the antibodies and destroys them. Live attenuated vaccines are the most popular type of vaccine. Table 1 shows their advantages and disadvantages.

Vaccination programmes and ethical issues

Mass vaccination breaks the chain of infection and makes it difficult for outbreaks of disease to occur and spread. The key to success is that most people are vaccinated (herd effect). However, the effect would soon disappear if the number of people vaccinated fell to levels where pathogens easily spread among unprotected individuals. This was demonstrated with concerns about the safety of the whooping cough vaccine in the 1970s and the MMR vaccine in the early 2000s. Vaccination rates dropped and instances of the diseases in the UK population rose. These scares show the difficulty of balancing individual well-being and freedom of choice with what is best for the majority. Keeping a sense of proportion in the light of factual information is essential if informed choices are to be made.

Despite success against smallpox, the antigenic variability of pathogens means that vaccination programmes rarely eliminate diseases. The annual programmes of 'flu vaccination highlight the problem. However international efforts to vaccinate whole populations means that diseases such as rubella (German measles), polio, measles, and typhoid are less common than before.

▼ **Table 1** *Live attenuated vaccines*

Advantages	Disadvantages
Low dose of vaccine possible as pathogens multiply in vaccine recipient.	Pathogen may mutate making vaccine ineffective or pathogen may revert to disease-catching form (uncommon).
Memory cells produced more effectively with live multiplying microorganisms.	Live vaccines must be stored in cool conditions.

Summary questions

1 What is a vaccine? [2]

2 Explain the difference between active immunity and passive immunity. [4]

3 What is the herd effect? [3]

5.7 The human immunodeficiency virus (HIV)

Specification reference: 3.2.4

> **Revision tip**
> Infection with HIV occurs when the virus passes from an infected person to someone else (transmission) in semen, blood, breast milk, or vaginal fluids.

> **Key terms**
> **Mass spectrometry:** Identifies what substances something (e.g. viruses) is made of.
> **Atomic force microscopy:** High-resolution scanning probe microscopy.
> **Virus load:** Amount of HIV virus in the blood of HIV positive people.

> **Synoptic link**
> For more information on viruses, see Topic 3.6, Prokaryotic cells and viruses. See also Topic 2.1, Structure of RNA and DNA.

> **Revision tip**
> Enzyme linked immunosorbent assay (ELISA) detects the presence (or not) of HIV antibodies in a sample of blood taken from the person being tested.

> **Summary questions**
> 1. Why is HIV called a retrovirus? [1]
> 2. Describe how HIV replicates. [4]
> 3. Identify *three* stages in the lifecycle of HIV that are targets for drug therapy. [3]

Structure

Mass spectrometry and atomic force microscopy are techniques used to investigate the structure of viruses.

Replication

Viruses cannot reproduce independently. To replicate viruses must enter living host cells. The HIV replication cycle is:

- HIV binds to CD4 protein receptors on the cell surface of the host T_H cells
- the HIV capsid containing viral RNA and different enzymes including reverse transcriptase enters the host cell
- the RNA reverse transcribes into a double-stranded DNA, catalysed by reverse transcriptase
- the double-stranded virus DNA becomes part of the DNA of the host cell, catalysed by the enzyme integrase.
- virus DNA may remain dormant – this is the latent stage of HIV infection
- when activated, the virus DNA transcribes mRNA which translates into new virus proteins
- the virus proteins assemble into new virus particles which are released from the host cell into the bloodstream to infect other T_H cells.

Treatment and control of HIV infection

Drugs aim to target different parts of the HIV replication cycle.

- **Receptor antagonists** block binding between HIV attachment proteins and CD4 host cell receptor protein.
- **Fusion inhibitors** prevent the outer envelope of HIV fusing with the cell-surface membrane of the host cell.
- **Reverse transcriptase inhibitors** prevent synthesis of virus DNA from virus RNA.
- **Integrase inhibitors** prevent virus DNA becoming part of host cell DNA.
- **Protease inhibitors** prevent assembly of virus proteins into new virus particles.

HIV positive people are treated with a cocktail of drugs (combination therapy). If some of the drugs in the cocktail are less effective because of mutations of the virus, then other drugs will continue to reduce the affected person's load of virus, reducing the risk of infection developing into AIDS.

Antibiotics

Antibiotics are used to treat diseases caused by bacteria. The drugs target different structures in bacteria, e.g. the cell wall of bacteria contains murein. Viruses are not surrounded by a cell wall of murein. There is nothing for antibiotics to target; therefore the drugs are ineffective against viruses.

Chapter 5 Practice questions

1. Antibodies consist of four polypeptide chains: two long heavy chains and two short light chains.

   a. Explain why antibodies have a quaternary structure. *(1 mark)*

   An antibody molecule consists of a constant region and variable region. The different regions are the result of the evolution of the molecule.

   b. In terms of selection pressure, suggest why the constant region is very similar in all types of antibody. *(2 marks)*

   c. Explain the relationship between the structure of the variable region and the specificity of an antibody to a particular antigen. *(3 marks)*

2. a. Distinguish between the innate immune system and adaptive immune system. *(3 marks)*

   The bacterium *Clostridium tetani* produces a toxin that disrupts the transmission of nerve impulses. If a deep wound is infected by the bacterium, the person concerned is at risk within three to four days of developing a condition where body muscles become permanently contracted. The condition is called tetanus. The graph shows a person's immune response to infection by the tetanus antigen. It takes up to two to three weeks for tetanus antibodies to be produced. Use the information, the graph, and your own knowledge to answer the following questions:

   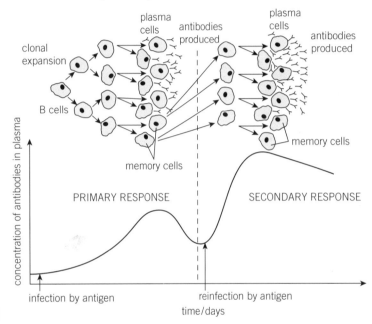

   b. Give *two* reasons why a person should have an anti-tetanus vaccination as soon as possible after possible infection by tetanus bacteria. *(2 marks)*

   c. If the person has not been vaccinated before possible exposure to infection by tetanus bacteria, explain why the vaccine given is usually a preparation of anti-tetanus antibodies, produced by another animal. *(2 marks)*

   d. State the type of immunity provided by a vaccine consisting of antibodies produced by another animal. *(1 mark)*

   e. Explain why the immunity provided by a vaccine consisting of antibodies produced by another animal is not long lasting. *(2 marks)*

   f. When exposed to an antigen for the first time, a person may develop symptoms of disease before recovering. When exposed to the antigen for a second time and subsequently, the same person does not develop symptoms of that disease. Using the graph in the figure, explain why. *(3 marks)*

3. a. The term retrovirus is used to describe HIV. Define the term. *(2 marks)*

   HIV replicates inside host cells. Release of newly replicated HIV destroys the host cells. T helper cells of the immune system are host cells to HIV.

   b. Describe how HIV enters T helper host cells. *(1 mark)*

   c. Explain why HIV infection of T helper cells weakens the immune system. *(1 mark)*

   Statement: "HIV does not directly cause the diseases associated with AIDS."

   d. Explain this statement. *(2 marks)*

6.1 Exchange between organisms and their environment
Specification reference: 3.3.1

key
SA = surface area
V = volume

cube A
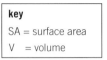

cube B

cube C
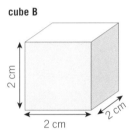

▲ Figure 1

Exchanging materials across a surface

All cells (tissues, organs, organisms) exchange gases, food, and other materials with their environment. These exchanges occur across the surfaces of cells. The larger the surface, the more material can be exchanged.

Surface area to volume ratio

Figure 1 and Table 1 show calculations of surface area (SA) and volume (V) for three cubes of different sizes. (Remember that a cube has six faces.)

▼ Table 1

	cube A	cube B	cube C
SA of one face	$1 \times 1 = 1\ cm^2$	$2 \times 2 = 4\ cm^2$	$3 \times 3 = 9\ cm^2$
SA of cube	$1\ cm^2 \times 6 = 6\ cm^2$	$4\ cm^2 \times 6 = 24\ cm^2$	$9\ cm^2 \times 6 = 54\ cm^2$
V of cube	$1 \times 1 \times 1 = 1\ cm^3$	$2 \times 2 \times 2 = 8\ cm^3$	$3 \times 3 \times 3 = 27\ cm^3$
$\frac{SA}{V}$	$\frac{6}{1} = 6$	$\frac{24}{8} = 3$	$\frac{54}{27} = 2$

Cells (tissues, organs, organisms) are not usually cubical, but the calculations apply to any shape. For example, as a cell grows it takes in more nutrients and gases and produces more waste substances.

After the cell reaches a certain size, its surface area becomes proportionately too small to take in enough of the substances it needs and remove enough of the wastes it produces. At this point the cell divides into two smaller daughter cells whose surface area to volume ratio is greater than that of the parent cell, enabling enough food and gases to pass into the cells and wastes to pass out of the cells.

Organ systems specialised for exchanging materials

All organisms exchange gases, food, and other materials between themselves and the environment. The exchanges take place across body surfaces. Changes in shape are adaptations that increase surface area, maximising the exchange of materials between an organism and its surroundings (Figure 2).

Revision tip
- The SA/V of cube B is half that of cube A.
- The SA/V of cube C is two-thirds that of cube B and one-third that of cube A.

Remember:
- The *larger* the cube becomes, the *smaller* its SA/V because SA increases more slowly than V.
- Surface area increases with the square (power^2) of the side.
- Volume increases with the cube (power^3) of the side.

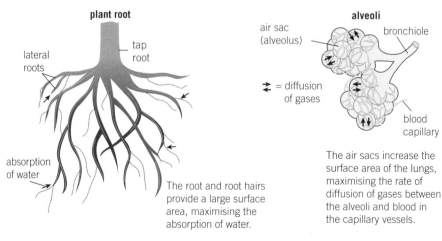

▲ Figure 2 *Changes in shape are adaptations which increase the surface area to volume ratios of these organisms*

Question and model answer

Q Why does a small mammal such as a vole (a rodent about 5 cm long) eat its own weight of food each day?

A *A vole's body has a large surface area relative to its volume. Its metabolic rate (rate of chemical reactions taking place in its cells) is high and much heat therefore is lost from its body through the skin to the environment. So, it needs lots of food as a source of energy to help maintain a constant body temperature.*

Summary questions

1. Explain why the surface area to volume ratio (SA/V) of large organisms is less than that of small organisms. [2]

2. Briefly explain why an increase in size might be a stimulus for a cell to divide. [2]

3. Explain why facing the sun on a hot day helps lizards to regulate their body temperature. [2]

6.2 Gas exchange in single-celled organisms and insects

Specification reference: 3.3.2

> **Synoptic link**
>
> Revise Topics 4.2, Diffusion, and 4.3, Osmosis, before starting on this topic.

Gas exchange in *Amoeba*

Amoeba proteus is a single-celled organism that lives in freshwater.

- The concentration of dissolved oxygen outside the cell is greater than the concentration of oxygen inside the cell. As a result oxygen diffuses down its concentration gradient from the water into the cell.
- The rate of diffusion of oxygen across the membrane is sufficient to satisfy the oxygen requirements of the cell. Carbon dioxide is produced by aerobic respiration inside the cell.
- The concentration of carbon dioxide inside the cell is greater than the concentration of carbon dioxide in solution outside. As a result carbon dioxide diffuses down its concentration gradient from the cell into the water.

Gas exchange across the tracheae of insects

Most insects live on land. The insect body, covered by an exoskeleton, is impermeable to water. This helps prevent water loss from the insect's body. However, this waterproofing means that the exoskeleton is also impermeable to oxygen and carbon dioxide from the air. Air enters and leaves the insect body through a set of pores opening at the body's surface.

- The openings are called **spiracles**.
- Tubes called **tracheae** and **tracheoles** extend from the spiracles into the body's tissues.

Rings of chitin (a carbohydrate) prevent collapse when the pressure of air inside the tracheae is less than atmospheric pressure.

The tracheal system at work

A solution of gases and other substances fills the ends of tracheoles. Oxygen and carbon dioxide are exchanged between the solution filling the ends of tracheoles and the cells of tissues nearby. As a result oxygen is supplied to the tissues; carbon dioxide is removed from the tissues.

At rest the water potential of the tissues is less negative than the solution in the tracheoles. If activity increases, respiration switches from aerobic to anaerobic.

- As a result lactic acid is produced in the cells of tissues.
- As a result the water potential of the tissues becomes more negative than the solution in the tracheoles.
- As a result water passes from the tracheoles to the tissues by osmosis.
- As a result more air is drawn into the tracheoles.
- As a result more oxygen is available to tissues.

When activity stops, the lactic acid is oxidised and the water potential in the tissues becomes less negative. As a result water passes from the tissues to the tracheoles by osmosis.

In larger insects contraction and relaxation of the abdominal muscles ventilates the tracheal system. As a result the flow of air through the system increases.

> **Summary questions**
>
> 1 Explain why concentration gradients of oxygen and carbon dioxide across the surface of the single-celled organism *Amoeba proteus* enable it to exchange the gases between itself and its environment. [2]
>
> 2 Explain why gas exchange does not take place across the exoskeleton of an insect. [1]
>
> 3 Explain how gas exchange takes place between the tracheae and tissues of an insect. [5]

6.3 Gas exchange in fish
Specification reference: 3.3.2

Gas exchange across the gills of bony fish

The gill slits are a series of openings on either side of the head of a fish. Each gill slit is separated by a thin, vertical bar of bone called the gill arch. Rows of leaf-like tissue called lamellae project from either side of the gill arch. Oxygen and carbon dioxide are exchanged across the lamellae between the blood flowing through them and the water flowing over them. The operculum is a flap which covers the gill slits of bony fish. Each lamella is folded into gill plates, increasing the surface area across which gases are exchanged. Vessels supplying the lamellae with blood branch into a dense capillary network within each lamella (see Figure 1).

Inspiration – lowering the floor of the pharynx *increases* the volume of the mouth cavity and reduces pressure. Water flows in through the open mouth and over the lamellae.

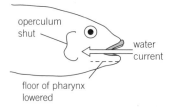

Expiration – raising the floor of the pharynx *reduces* the volume of the mouth cavity and increases pressure. The **pressure pump** effect pushes water over the lamellae and against the opercula, pressing them open.

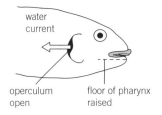

▲ Figure 2

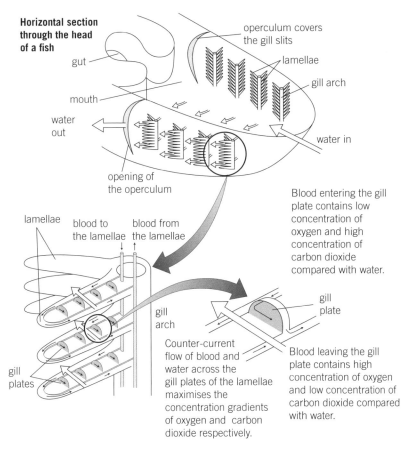

▲ Figure 1 *The gills of a fish are made up of gill arches, lamellae, and gill plates across which gas exchange takes place*

The lamellae at work

The concentration of oxygen in water is greater than in the blood passing to the gills. Oxygen diffuses from the water into the blood. The concentration of carbon dioxide in the blood passing to the gills is greater than in the water. Carbon dioxide diffuses from blood into the water. The concentration gradients of gases between blood and water are maximised because the direction of blood flow through the lamellae is opposite to that of the water flowing over the lamellae; the so-called counter-current effect.

Flow of water

The flow of water through the mouth and over the lamellae is called the respiratory current. The opening and shutting of the mouth and opercula are coordinated. There is a continuous flow of water over the lamellae.

Summary questions

1. Describe adaptations of the gills of a fish which maximise their surface area. [2]

2. Explain how the counter-current effect maximises the exchange of oxygen and carbon dioxide in solution between water and the blood flowing through the gill lamellae of a fish. [4]

3. Explain how the pressure pump effect promotes the flow of water over the gill lamellae of a fish. [5]

6.4 Gas exchange in the leaf of a plant

Specification reference: 3.3.2

Gas exchange across the leaves of plants

Figure 1 shows how different adaptations maximise the exchange of gases between the leaves of a plant and the environment.

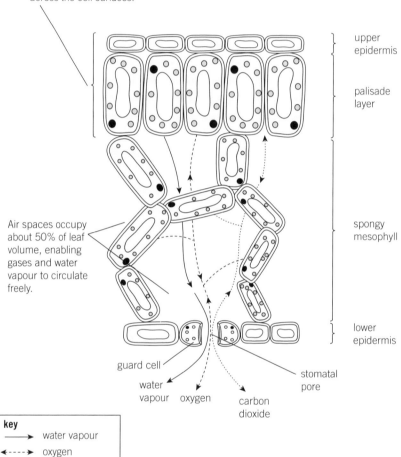

▲ **Figure 1** *The gas exchange surfaces inside the leaf of a dicotyledonous plant*

The under surface of the leaf is perforated with gaps called stomata. Each stoma (singular) is flanked by guard cells which contain chloroplasts. The direction of the net movement of oxygen and carbon dioxide molecules between the inside of a leaf and the atmosphere depends on the balance between the rate of photosynthesis and the rate of aerobic respiration of the leaf's tissues. When the concentration of carbon dioxide produced in aerobic respiration balances that used in photosynthesis, then the net exchange of carbon dioxide between the leaf and the atmosphere is zero. This is the compensation point.

The spaces inside a leaf are saturated with water vapour and the concentration of water vapour in the atmosphere is usually less than inside the leaf. As a result water vapour passes down its concentration gradient from inside the leaf, through the stomata perforating the under surface of the leaf, to the atmosphere. The process is called transpiration.

Summary questions

1. Explain the role of air spaces in gaseous exchange in the leaf of a plant. [1]

2. Explain why the net exchange of oxygen and carbon dioxide between a leaf and the atmosphere is zero at the compensation point. [2]

3. What is transpiration? Explain the process. [4]

6.5 Limiting water loss

Specification reference: 3.3.2

How stomata open and close

When guard cells fill with water and become more turgid, their volume increases. As a result the guard cells push each other apart and the stomatal pore opens. Guard cells become more turgid because of the active transport of potassium ions (K^+) into the cells from the surrounding cells of the leaf's lower epidermis. As a result the water potential of the guard cells becomes more negative compared with the cells surrounding them, water passes into the guard cells from the surrounding cells by osmosis, and the guard cells become turgid and bow outwards, opening the stomatal pore.

The active transport of potassium ions is triggered by light and photosynthesis. In the dark, the active transport of ions stops. As a result the ions diffuse down their respective concentration gradients until equilibrium with the surrounding epidermal cells is reached. The water potential of the guard cells becomes less negative compared with the cells surrounding, the water passes from the guard cells into the surrounding cells by osmosis, the guard cells become less turgid, and the pore closes.

Cacti are xerophytic plants. Their adaptations are a compromise between the opposing requirements of gas exchange and the limitation of the loss of water from the plant to the environment.

- Leaves are pointed spines, reducing their surface area and so reducing water loss.
- A thick waxy waterproof layer covering the surfaces also reduces water loss.
- A shiny surface reflects heat away from the cactus surface.
- Storage of water in the thick stem enables the cactus to survive long periods without rain.
- The large surface area of the roots enables the cactus to maximise absorption of water in short supply.

Marram grass grows on sand dunes. Its root system spreads out deep within the sand dune. In dry conditions the leaves of marram grass roll up putting the stomata on the inside of the tube. The stomata are sunk in pits and fringed with hair-like extensions of the lower leaf surface. As a result the moist air is trapped within the tube against the lower leaf surface, the water potential gradient between the inside of the leaf and the humid air trapped within the tube is minimised, and the loss of water from the plant is minimised.

> **Revision tip**
> Water vapour passes down its concentration gradient from inside the leaf, through its stomata, to the atmosphere. Control of the opening and closing of stomata limits the loss of water from the plant.

> **Revision tip**
> The part of the wall of each guard cell facing the opening of the stomatal pore is thicker and less flexible than the rest of the wall of each cell. When turgid, the guard cells swell, the cell walls stretch unevenly, so the cells bow outwards as the pore opens.

> **Key term**
> **Xerophytic:** (Of plants) able to survive in hot, dry conditions.

> **Key term**
> **Adaptation:** Characteristic which increases an organism's chances of surviving in its habitat.

Summary questions

1 Summarise the process which controls the opening and closing of the stomata of a leaf. [4]

2 What is an adaptation? Identify the similar adaptations that reduce the loss of water from cacti and marram grass. [5]

6.6 Structure of the human gas-exchange system

Specification reference: 3.3.2

In Figure 1:

- the trachea (windpipe) branches into two bronchi
- each bronchus branches into many bronchioles
- each bronchiole ends in a cluster of alveoli (air sacs)
- the alveoli honeycomb the lung tissue.

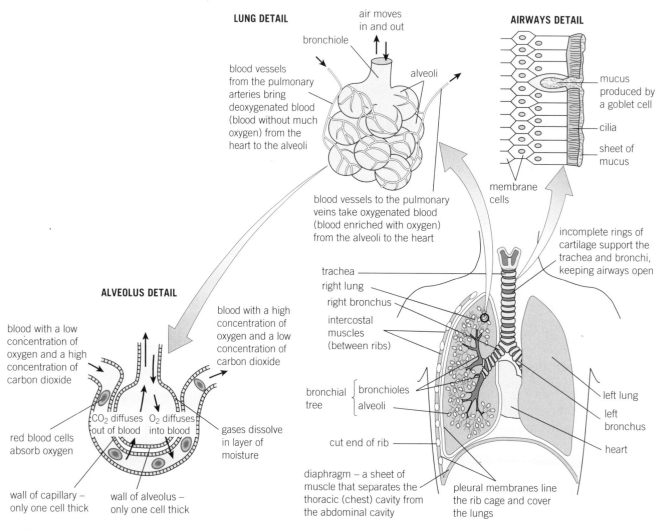

▲ **Figure 1** *The structures of the human gas-exchange system*

A network of capillary vessels supplies blood to and carries blood from the alveoli. The walls of the alveoli and capillary vessels each consist of squamous epithelium one cell thick. As a result a surface only two cells thick separates the air in the alveolus and the blood in the capillary blood vessel, and the diffusion of gases (oxygen and carbon dioxide) across the surface is rapid because the diffusion pathway is short.

The rate of diffusion of gases across the surface of the alveoli depends on concentration gradients. Inhalation (breathing in) draws air into the alveoli. The concentration of oxygen in inhaled air in the alveoli is greater than the concentration of oxygen in the blood supplied to the alveoli. As a result oxygen diffuses down its concentration gradient from the air in the alveoli to the blood in the capillary vessels supplying the alveoli.

The concentration of carbon dioxide in the blood supplied to the alveoli is greater than that in the inhaled air in the alveoli. As a result carbon dioxide diffuses down its concentration gradient from the blood in the capillary vessels supplying the alveoli to the air in the alveoli.

Exhalation (breathing out) carries air out of the lungs. Table 1 shows the percentage change in oxygen and carbon dioxide of inhaled and exhaled air.

▼ **Table 1** *Adaptations enable the human lung to exchange gases efficiently*

Structure	Adaptations and functions
Trachea and bronchi	Provide route for air to enter and leave the lungs. Ventilation maintains the concentration gradient of oxygen and carbon dioxide in the alveoli, maximising the rate of diffusion. Mucus-secreting cells (goblet cells) and ciliated epithelium trap and remove bacteria, viruses, and other particles from the incoming air. Cartilage rings support the airways and prevent them collapsing as air pressure in the lungs changes during ventilation.
Bronchioles	Finely branching tubes that end in alveoli. At the end bronchioles have no cartilage rings. Smooth muscle in the bronchioles enables control of diameter, e.g. adrenaline relaxes the muscles increasing the diameter of the bronchioles allowing better air flow during exercise.
Alveoli	Very numerous providing a large surface area that maximises the rate of diffusion. Walls are one cell thick, and in close contact with capillaries, also with walls one cell thick. The diffusion pathway of oxygen from the air in the alveoli and of carbon dioxide in the blood of the capillary blood vessels is therefore short. Elastic fibres enable alveoli to recoil (spring back) after expiration and to extend during inspiration.

Summary questions

1. Summarise the features of the alveolar epithelium as a gas-exchange surface. [4]

2. Explain how the diffusion of oxygen and carbon dioxide across the surface of the alveoli depends on their respective concentration gradients. [4]

6.7 The mechanism of breathing

Specification reference: 3.3.2

Key term

Pulmonary ventilation: Breathing in and out.

The mechanism of breathing

Pulmonary ventilation is the result of movements of the cage around the thoracic (chest) cavity formed by the ribs and diaphragm. The cage is elastic and as it moves the pressure of air in the lungs changes. The change in air pressure causes inhalation (breathing in) and exhalation (breathing out). These repeated breathing movements ventilate the lungs. The pulmonary ventilation rate (breathing rate) is the number of inhalations / exhalations per unit time.

Inhalation

Refer to Figure 1.

- The diaphragm contracts and becomes less dome-shaped.
- At the same time the intercostal muscles between the ribs contract and raise the rib cage.
- The thoracic cavity enlarges. The resulting reduction in air pressure is transmitted via the pleural cavity to the lungs.
- The pressure of air in the alveoli is less than that of the atmosphere. Air, therefore, is drawn into the lungs through the trachea and bronchi.

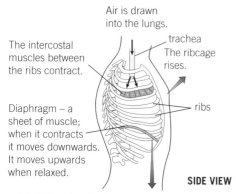

Inhalation: the diaphragm contracts and flattens.

▲ Figure 1

Exhalation

Refer to Figure 2.

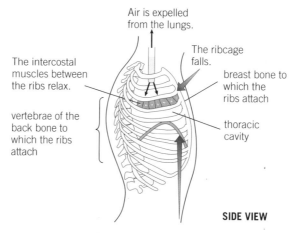

The diaphragm relaxes and curves upwards.

▲ Figure 2

- The diaphragm and intercostal muscles relax, lowering the ribs and raising the diaphragm.
- The volume of the thoracic cavity decreases and the lungs are compressed.
- The pressure of air in the alveoli is greater than that of the atmosphere.
- The air passes from the lungs through the bronchi and trachea to the atmosphere.

The pleural cavity is the space between the lungs and the rib cage. It is lined by pleural membranes. The membranes are lubricated, facilitating (making easy) movements of the lungs. Gas pressure within the pleural cavity is less than that of the atmosphere.

Summary questions

1. Summarise how air pressure changes in the lungs during one cycle of inhalation and exhalation. [4]

2. Distinguish between the action of the internal intercostal muscles and external intercostal muscles during one cycle of inhalation and exhalation. [3]

6.8 Exchange of gases in the lungs
Specification reference: 3.3.2

Changes in the volume of inhaled and exhaled air

The volume of air exchanged between the lungs and the atmosphere depends on the body's activity and the extent and rate of ventilation (measured as the breathing rate) of the lungs. The volume of air breathed in and out is measured using a spirometer.

The air capacity of the human lungs is about 5.5 dm^3, of which 1.5 dm^3 is the residual volume. The term refers to the air not removed from the lungs even when breathing is forced.

At rest, normal breathing results in a tidal volume of around 0.5 dm^3. So a person with a resting breathing rate of 16 inhalations/exhalations per minute exchanges air at a rate of $16 \times 0.5 = 8$ dm^3 min^{-1}.

Physical activity increases the breathing rate. The whole volume of the lungs less its residual volume comes into play. The volume is about 4.0 dm^3 (5.5 dm^3 – 1.5 dm^3) and represents the vital capacity of the lungs. For example, a breathing rate of 40 inhalations/exhalations per minute exchanges air at a rate of $40 \times 4.0 = 160$ dm^3 min^{-1}.

Conditions affecting the exchange of gases in the lungs

Asthma is a common condition that is caused by obstruction to the flow of air through the airways. Difficulty with breathing is a common symptom. Obstruction to the flow of air is due to the inflammation of the airway walls and their subsequent narrowing. Lung disease may also destroy lung tissue, reducing gas exchange between air in the alveoli and in the blood.

Smoking and lung disease

Research begun in the 1930s gathered data which by the 1970s established the link between smoking, lung cancer, and other diseases.

- Doctors quickly realised the possible significance of the data. Many gave up smoking. Deaths from lung cancer among doctors went down compared with the population as a whole, who were less well informed.
- Further studies established the correlation between the risk of dying from lung cancer and the number of cigarettes smoked – the more cigarettes smoked, the greater the risk.

Non-smokers also suffer increased risks of ill health when they breathe in smoke from other people's cigarettes: 'passive' smoking. The evidence supports the idea that people have the right to a smoke-free environment. A change in UK law has made public transport and places where people work, shop, and find entertainment smoke-free zones.

Pollution and lung disease

Air is polluted when substances generated by human activities are released into the atmosphere. For example nitrogen oxides are formed when we burn fossil fuels in power stations and engines. These pollutants are linked to different lung diseases.

Revision tip
Narrowing of the airway walls during an asthma attack is the result of contraction of the smooth muscle of the walls.

Key term
Pulmonary vascular resistance (PVR): The extent to which blood flow is impeded through the pulmonary circulation.

Revision tip
When something changes and another thing changes at the same time, this does not mean that the something that changes is the *cause* of the other thing changing. We say that the changes are **correlated**. To prove *cause*, statistical tests are used to analyse many samples/examples of the correlation. If the tests make it overwhelmingly likely that one change causes another change then it would be perverse (unreasonably obstinate) to deny cause and effect. Make sure you can distinguish between correlation and cause and effect.

Summary questions

1 Explain the relationship between residual volume, tidal volume, and vital capacity of the lungs. [2]

2 Fibrosis refers to the thickening and inflexibility of lung tissue. What is the difference between fibrosis and asthma? [3]

3 Why do people suffering from emphysema quickly become breathless and exhausted? [1]

6.9 Enzymes and digestion
Specification reference: 3.3.3

> **Synoptic link**
> It will help you understand this topic if you revisit Topic 1.3, Carbohydrates: disaccharides and polysaccharides, Topic 1.5, Lipids, Topic 1.6, Proteins, and Topic 1.7, Enzyme action.

> **Revision tip**
> The gut is a muscular tube running from the mouth to the anus. Along it, different physical and chemical processes break down food into substances suitable for absorption into the blood stream.

> **Revision tip**
> The human gut is 7–9 m long. The longest part of it consists of the small intestine and large intestine. These lie folded and packed into the space of the abdominal cavity. The small intestine (more than 5 m long!) has a smaller diameter than the large intestine.

> **Revision tip**
> The ending –*ase* refers to a category of enzymes or a particular enzyme. For example *proteases* (category); *amylase* (enzyme).

> **Revision tip**
> Enzymes as protein components of the cell-surface membrane complete the digestion of disaccarides → monosaccharides.

Digestion: basic principles
Our body needs the nutrients (i.e. carbohydrate, lipid, and protein) that food contains. They are mostly complex insoluble polymers which the body cannot absorb. The processes of digestion break down the polymers into their soluble constituent units which the body can absorb. Breakdown occurs by hydrolysis. Mineral ions and vitamins are also nutrients which are small and soluble and not digested, but are absorbed unchanged into the blood stream.

Different enzymes catalyse the hydrolytic reactions of digestion grouped according to the reactions they catalyse:

- **Carbohydrases** catalyse the digestion of carbohydrates such as starch into simple sugars.
- **Proteases** catalyse the digestion of proteins into amino acids.
- **Lipases** catalyse the digestion of triglycerides into fatty acids and glycerol.

Carbohydrate digestion
The action of amylase
Saliva, produced from salivary glands, pours onto food in the mouth. Saliva contains:

- salivary amylase, the carbohydrase which catalyses the hydrolysis of starch to maltose
- chloride ions (Cl^-), which activate amylase.

Pancreatic juice, produced by the pancreas, passes from the pancreas through the pancreatic duct into the duodenum. The juice contains amylase, which continues to catalyse the digestion of starch.

Disaccharidases produced in the intestinal epithelium
Disaccharidases are secreted by the epithelial cells of the intestinal wall.

The disaccharidases include:

- **maltase** which catalyses the hydrolysis of maltose to glucose
- **sucrase** which catalyses the hydrolysis of sucrose to glucose and fructose
- **lactase** which catalyses the hydrolysis of lactose to glucose and galactose. The absence of lactase is the cause of **lactose intolerance** in affected individuals.

Protein digestion
- **Endopeptidases** in the gastric juice and pancreatic juice, and secreted by the cells of the small intestine, catalyse hydrolysis of the peptide bonds within protein molecules. As a result proteins are broken down into smaller molecules of peptides and polypeptides.
- **Exopeptidases** in the pancreatic juice and secreted by the cells of the small intestine catalyse hydrolysis of the peptide bonds linking the amino acids at each end of peptide/polypeptide chains. As a result amino acids are released and the peptide/polypeptide chains progressively shorten until only *dipeptides* remain.
- **Dipeptidases** catalyse hydrolysis of the peptide bond joining the pairs of amino acids forming dipeptides. As a result amino acids are released.

Lipid digestion

Lipases catalyse hydrolysis of the ester bonds of some (but not all) triglycerides following emulsification of the lipids by bile salts. Lipases catalyse a type of reaction (hydrolysis of ester bonds) rather than a particular reaction. The lack of specificity means that lipid digestion, unlike protein digestion, does not follow a precise sequence of enzyme-regulated hydrolyses.

Emulsification of lipids

Bile juice is produced by the liver and stored in the gall duct before release into the duodenum through the bile duct. Bile salts (potassium taurocholate and sodium taurocholate) in the bile juice emulsify lipids. Emulsification breaks up lipids into tiny droplets, increasing the surface area on which lipases can react. As a result the rate of lipid digestion increases.

Summary questions

1. Distinguish between a carbohydrase and amylase. [2]
2. Name the products of digestion of a triglyceride. [2]
3. Distinguish between the action of endopeptidases and exopeptidases. [2]

6.10 Absorption of the products of digestion

Specification reference: 3.3.3

> **Synoptic link**
>
> You will better understand this topic if you look back at Topic 4.2, Diffusion and Topic 4.5, Co-transport and absorption of glucose in the ileum, and also read Topic 7.6, Blood vessels and their functions.

Surface area

The inner surface of the wall of the small intestine has folds, villi, and other features which greatly increase the surface area and therefore the rate of absorption of the products of digestion. The surface of the cells of the epithelium lining the villi is thrown into tiny folds, visible only in the electron microscope. The folds are called microvilli.

Absorption of monosaccharides and amino acids

Absorption of the products of carbohydrate and protein digestion (monosaccharides and amino acids) begins in the duodenum, but most occurs across the wall of the ileum. It starts at the microvilli of the epithelial cells at the surface of each villus.

Molecules of monosaccharides and amino acids pass

- through the plasma membrane of each epithelial cell
- across each cell
- through the plasma membrane on the opposite side of each cell
- through the plasma membrane of the cells of the wall of a capillary blood vessel.

A network of capillary blood vessels and lacteal vessels within each villus run into the hepatic portal vein and into the lymphatic system respectively. Passage of molecules of monosaccharides and amino acids across the plasma membrane of the cells is by diffusion, facilitated diffusion, and active transport.

Monosaccharides and amino acids pass into the hepatic portal vein to be transported in the bloodstream to the liver.

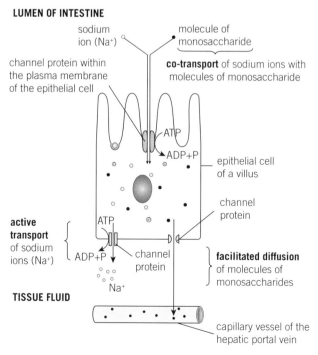

▲ Figure 1 *Absorption in the intestine*

Carrier proteins in the plasma membrane couple the transport of monosaccharides or amino acids (depending on the type of protein channel involved) with sodium ions (Na^+) from the lumen of the intestine into the epithelial cells of the villi. The process is called co-transport because the movement of monosaccharides/amino acids and sodium occurs together. The process is a form of active transport requiring ATP as a source of energy. Monosaccharide, amino acid, glycerol, and some water-soluble fatty acid molecules transfer from the epithelial cells into capillary vessels of the hepatic portal vein by facilitated diffusion. Sodium ions (Na^+) are pumped from the base of epithelial cells into the surrounding tissue fluid by active transport. ATP is required as a source of energy.

Absorption of glycerides and fatty acids

- Undigested triglycerides and monoglycerides, and most fatty acids, combine with bile salts that emulsified lipid droplets in the first place, forming tiny (4–7 nm diameter) **micelles.**
- Movement of the contents in the lumen of the intestine delivers micelles to the surface of the epithelial cells of the villi.
- At the surface, micelles break down, releasing their contents which are absorbed into the villus cells by diffusion.
- Vitamins A and D and cholesterol are also transported to the surface of villus cells in micelles.
- Within the villus cells triglycerides are resynthesised on the smooth endoplasmic reticulum from fatty acids and monoglycerides, and then transferred to the Golgi apparatus.
- Here the molecules are packaged in an envelope of protein, cholesterol, and phospholipid.
- The packages are called **chylomicrons**. They pass out of the opposite side of the cells of the villus into the tissue fluid by **exocytosis**, and enter the lacteal vessels.
- Chylomicrons circulate in the lymph of the lymphatic vessels and drain into the **left subclavian vein** via the **thoracic duct** in the neck.

Summary questions

1 Briefly explain how molecules of monosaccharides and amino acids are absorbed by a cell lining the wall of the ileum (small intestine). [4]

2 Describe the route by which molecules of monoglycerides and most fatty acids pass from the lumen of the ileum (small intestine) to the blood. [6]

Go further

a Adaptations of the cells of the surface of the small intestine and the surface itself maximise the rate of absorption of the products of digestion.

  i State how these adaptations increase the rate of absorption of the products of digestion.

  ii Distinguish between the adaptation of the cells of the surface of the small intestine and the surface itself increasing the rate of absorption of the products of digestion.

b Discuss the role of the proteins of the surface membrane of the cells lining the small intestine involved in the absorption of fructose, glucose, and amino acids.

c Triglycerides are hydrolysed to molecules of glycerol and fatty acids but reform within villus cells and are then absorbed by the lacteal vessels of the lymphatic system. Explain how.

Chapter 6 Practice questions

1 The diagram represents two alveoli and a blood capillary in lung tissue.

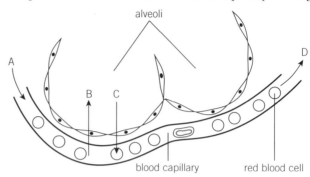

    a Which arrow represents:
        i deoxygenated blood
        ii carbon dioxide? *(2 marks)*
    b i Explain the process by which oxygen passes from the air in the alveoli to the red blood cells in the capillary. *(2 marks)*
        ii Through how many layers of cells does the oxygen pass? *(1 mark)*

The walls of the alveoli consist of a layer of cells called squamous epithelium. The cells are adapted in a way that maximises the rate of exchange of gases between blood in the blood capillary vessel and air in the alveoli.

    c Describe the appearance of the cells forming the layer of squamous epithelium. *(1 mark)*
    d What is an adaptation? *(1 mark)*
    e Explain how the adaptation of squamous epithelium maximises the rate of gaseous exchange between blood and air. *(2 marks)*

2 a Gas exchange occurs across an earthworm's moist body surface between its tissues and the atmosphere. Gas exchange cannot occur across the body surface of an insect of similar mass and surface area. Explain why. *(2 marks)*

The figure represents part of the gas exchange system of an insect:

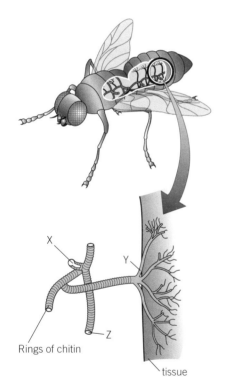

    b i Name the structures labelled X, Y, and Z. *(3 marks)*
        ii Explain the role of the rings of chitin. *(2 marks)*

The wing muscles of a flying insect contract and relax vigorously. Lactate (lactic acid) is produced and accumulates in the muscle tissue.

    c Explain why lactate accumulates in the tissue of the wing muscles of a flying insect. *(2 marks)*
    d What is the effect of the accumulation of lactate on the water potential of the muscle tissue? *(2 marks)*
    e Explain the relationship between the water potential of the insect muscle tissue and the supply of oxygen to the tissue. *(3 marks)*

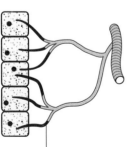

Cells of tissues at rest are respiring aerobically. A solution of gases and other substances fills the ends of the tracheoles.

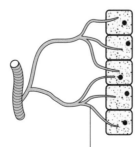

Cells of tissues are active and respiring anaerobically. Air fills the ends of the tracheoles.

7.1 Haemoglobin

Specification reference: 3.3.4.1

The haemoglobin molecule

Haemoglobin is the oxygen-carrying pigment in red blood cells. Its molecule is a globular protein with a quaternary structure, consisting of:

- four coiled polypeptide chains (the globin part of the molecule)
- four haem groups.

Each haem group is a prosthetic group; part of the protein but not made of amino acids. The four haem groups each contain an iron ion that combines with a molecule of oxygen. As a result a molecule of haemoglobin can combine with four molecules of oxygen.

How do the loading and unloading properties of haemoglobin help with oxygen transport?

The loading of oxygen refers to the uptake of oxygen by haemoglobin. The unloading of oxygen refers to the release of oxygen by haemoglobin.

The partial pressure of oxygen in the lungs is about 12 kPa. In the tissues the partial pressure of oxygen is much lower, 2 kPa in muscle tissue for example. Haemoglobin exposed to this concentration of oxygen is less than 20% saturated. The oxygen released from the haemoglobin molecules passes into the blood plasma, diffuses into the muscle cells, and is used in aerobic respiration.

The Bohr effect

- Reducing the partial pressure of carbon dioxide increases the oxygen load of the blood for a given partial pressure of oxygen.
- Increasing the partial pressure of carbon dioxide reduces the oxygen load of the blood for a given partial pressure of oxygen.
- Increasing temperature has a similar effect.

> **Revision tip**
> Most enzymes, antibodies, and some hormones are globular proteins made of polypeptide chains that fold into a spherical shape. They are usually soluble in water.

> **Revision tip**
> As well as transporting oxygen to the respiring cells, haemoglobin also carries about 10% of the waste carbon dioxide that is transported by the blood to the alveoli. Carbon dioxide combines with the globin part of the molecule to form carbaminohaemoglobin.

Summary questions

1 Why is haemoglobin described as a globular protein with a quaternary structure? [2]

2 Why is the haem group of a haemoglobin molecule referred to as a prosthetic group? [2]

3 What is the role of the haem group of a haemoglobin molecule? [1]

7.2 Transport of oxygen by haemoglobin
Specification reference: 3.3.4.1

> **Revision tip**
> - Blood which contains a lot of oxyhaemoglobin is called oxygenated blood and is bright red.
> - Blood with less oxyhaemoglobin in it is called deoxygenated blood and looks red/purple.

Transport of oxygen by haemoglobin

Oxygen combines with haemoglobin forming oxyhaemoglobin in tissues where the concentration of oxygen is high. It quickly releases its oxygen in tissues where the concentration of oxygen is low. Haemoglobin is therefore ideal for the transport of oxygen from the lungs to the rest of the body's tissues.

Uptake of oxygen and the dissociation curve

The partial pressure (in kPa) of a gas is a measure of its concentration. It is proportional to its percentage by volume in a mixture of gases.

- The atmosphere contains nearly 21% oxygen. The partial pressure of oxygen is therefore about 21 kPa.
- The combination of haemoglobin with oxygen depends on the partial pressure of oxygen in contact with it (see Figure 1).

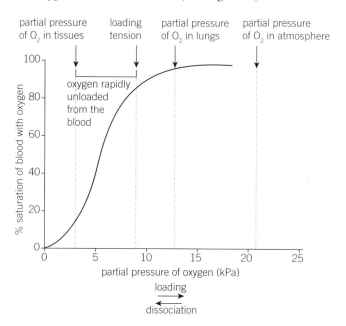

▲ **Figure 1** *Oxygen association/dissociation curve for adult haemoglobin*

Reading the graph in Figure 1 from left to right shows the relationship between the affinity (uptake) of oxygen by haemoglobin (association) and the increasing partial pressure (concentration) of oxygen in contact with it.

Haemoglobin takes up oxygen rapidly (has high affinity) for a relatively small increase in the partial pressure of the gas. The term loading tension refers to the point when 95% of the pigment is saturated. The loading tension corresponds to a partial pressure which is considerably less than the partial pressure of oxygen in the atmosphere. As a result the blood supplying the lungs becomes rapidly loaded with oxygen at the partial pressure of oxygen normally found in the lungs.

Reading the graph in Figure 1 from right to left shows the relationship between the release of oxygen by haemoglobin (dissociation) and the decreasing partial pressure of oxygen in contact with it. Haemoglobin releases (unloads) oxygen rapidly for a relatively small decrease in partial pressure of the gas. The unloading of oxygen corresponds to a partial pressure of oxygen normally found in tissues which are using oxygen in aerobic respiration. As a result tissues receive enough oxygen for their activities.

What makes the dissociation curve S-shaped? Remember that a haemoglobin molecule combines with four molecules of oxygen.

- The combination of oxygen with one haem group slightly changes the shape of the haemoglobin molecule.
- The shape change makes it easier for one of the other haem groups to load an oxygen molecule.
- The combination of oxygen with this haem group makes it even easier for a third haem group to load oxygen, and so on.

The S-shape of the dissociation curve is the result of these knock-on effects.

The effects work in reverse:

- As the partial pressure of oxygen decreases, an oxygen molecule may be released.
- Its loss slightly changes the shape of the haemoglobin molecule, making the unloading of subsequent oxygen molecules increasingly easy.

The alterations in the oxygen dissociation curve arising from the Bohr effect help to adjust the amount of oxygen tissues receive. For example, the demand for oxygen by muscle tissues during vigorous exercise is high. As the tissues respire, more and more carbon dioxide is released. The temperature of the tissues also increases. These changes in the micro-environment of the tissues cause the dissociation curve to shift to the right, increasing the supply of oxygen to the tissues at a time when it is needed.

Different types of haemoglobin

Vertebrates (fish, amphibia, reptiles, birds, and mammals), some insects, and many types of worm have different types of haemoglobin in their blood. Differences between the various molecules include variations in:

- the structure of polypeptide chains
- the number of polypeptide chains
- the number of haem groups.

The lugworm *Arenicola* lives in mud burrows in the intertidal zone of the seashore. The concentration of oxygen in its burrow is low between tides. The dissociation curve of its haemoglobin is shifted to the left. This means that the haemoglobin will still load oxygen even at low concentration. It only gives up oxygen when concentrations drop to very low levels, enabling *Arenicola* to survive when its burrow is uncovered at low tide.

> **Key term**
>
> **Allostery:** The changes in shape of a haemoglobin molecule as it successively combines with oxygen molecules. This is why we say that haemoglobin is an allosteric protein.

> **Synoptic link**
>
> Look back to Topic 7.1, Haemoglobin, for information on the Bohr effect.

Summary questions

1. When does the oxygen dissociation curve become an oxygen association curve? [2]

2. How are different types of haemoglobin suited to particular ways of life? [2]

7.3 Circulatory system of a mammal
Specification reference: 3.3.4.1

> **Synoptic link**
>
> A useful starting point to this topic would be to revise diffusion in Topic 4.2, Diffusion, and surface area to volume ratios in Topic 6.1, Exchange between organisms and their environment.

Mass transport

Some substances are needed by organisms and obtained from the environment. Other substances are wastes produced by organisms and removed to the environment.

In very small organisms where the ratio of surface area/volume is relatively large, substances can be exchanged between the organism and the environment by diffusion. In large organisms where the ratio of surface area to volume is relatively small, diffusion is too slow to meet their needs. Different systems of mass transport are needed to move substances rapidly from one part of a large organism to another.

The systems link with exchange surfaces where differences in concentration of substances and in pressure of gases propel substances:

- from where they are exchanged with the environment
- to where they are needed by tissues.

The circulatory (blood) system of mammals and the xylem and phloem tissues of plants are examples of mass transport systems.

The human circulatory system

The heart pumps blood through tubular blood vessels (arteries, veins, and capillaries). Blood transports oxygen, digested food, hormones, and other substances to the tissues and organs of the body. It also carries carbon dioxide and other waste substances (e.g. urea) from the tissues and organs of the body to where they are removed from the body.

Double and single circulatory systems

The human (mammalian) circulatory system is a double circulatory system. Blood flows through the heart twice for each complete circuit of the body. The pulmonary system (from the heart to the lungs and back) is separated from the systemic system (from the heart to the rest of the body and back). The two separate circuits allow rapid high-pressure distribution of oxygen in endothermic animals which have a relatively high metabolic rate.

Fish have a single circulatory system. Blood flows through the heart once for each complete circuit of the body. Fish have a relatively low metabolic rate. Blood leaving the gills is at a lower pressure as it has flowed through the exchange organs with their many branching capillaries.

Open and closed circulatory systems

The circulatory systems of mammals and fish are closed systems; the blood is contained in blood vessels. Insects have a mass transport system that is an open system. A fluid called hemolymph flows through the body in direct contact with tissues. Contractions of an open-ended, tubular heart circulate the hemolymph through the body.

> **Summary questions**
>
> 1. Explain why describing arteries as blood vessels that carry oxygenated blood or veins as blood vessels that carry deoxygenated blood is not a satisfactory definition of the pulmonary arteries and pulmonary veins respectively. [2]
>
> 2. Explain the role of the hepatic portal vein. [1]
>
> 3. a Describe the differences between a double circulatory system and a single circulatory system. [3]
>   b Explain the advantages of a double circulatory system. [3]

7.4 The structure of the heart

Specification reference: 3.3.4.1

Heart structure

The heart lies in the chest cavity, protected by the rib cage. The vessels bringing blood to and from the heart are the:

- **Aorta:** has arteries branching from it which carry oxygenated blood to the organs and tissues of the body, except the lungs. Contraction of the left ventricle pumps blood into the aorta.
- **Coronary arteries:** run over and deep into the walls of the heart. The arteries carry blood with supplies of dissolved oxygen and nutrients needed by the heart muscles. They branch from the aorta soon after it leaves the heart.
- **Pulmonary arteries:** carry deoxygenated blood *from* the heart *to* the lungs. Contraction of the right ventricle pumps blood into the pulmonary arteries.
- **Pulmonary veins:** carry oxygenated blood *to* the heart *from* the lungs.
- **Superior:** ('above' the heart in humans) **vena cava** carries deoxygenated blood from the head to the right atrium.
- **Inferior:** ('below' the heart in humans) **vena cava** carries deoxygenated blood from the rest of the body to the right atrium.

Inside the heart, the tricuspid and bicuspid valves (atrioventricular valves) separate the atria from their respective ventricles; the septum separates the right side of the heart from the left side.

The heart has two pumps

The right ventricle pumps blood through the pulmonary arteries and the left ventricle pumps blood through the aorta and through the arteries branching from the aorta.

Question and model answer

Q Why does the heart have two pumps?

A *As blood passes from the pulmonary arteries through the network of capillary blood vessels supplying blood to lung tissue to the pulmonary veins, there is a large drop in pressure. The pulmonary veins return the blood to the heart at low pressure. Contractions of the left ventricle increase blood pressure once more, generating enough force to propel the blood to all the other tissues and organs of the body.*

Question and model answer

Q Why is the wall of the left ventricle thicker than the wall of the right ventricle?

A *The distance that blood pumped from the left ventricle travels (through the aorta and the arteries branching from it) is much greater than the distance blood pumped from the right ventricle travels (through the pulmonary arteries to the lungs). The thicker wall of the left ventricle enables the ventricles to contract more powerfully with enough force to propel the blood the greater distance.*

Revision tip

The right side of the heart contains deoxygenated blood and the left side oxygenated.

Revision tip

It may seem odd that the heart needs its own blood supply, when its chambers are filled with blood. But its walls are so thick that oxygen and nutrients in the blood inside the heart would not be able to diffuse into all of the heart muscle.

Key terms

Renal arteries: Transport oxygenated blood to the kidneys.

Renal veins: Drain deoxygenated blood from the kidneys.

Summary questions

1 **a** Which blood vessels supply blood to the surface of and deep into the wall of the heart?
   b Why does the heart need its own blood supply? [2]

2 Explain how valves within the heart direct the flow of blood through the heart. [3]

3 Explain why the wall of the left ventricle is thicker than the wall of the right ventricle. [4]

7.5 The cardiac cycle

Specification reference: 3.3.4.1

> **Revision tip**
> The opening and closing of the different valves depend on the relative pressure on either side of each one.

The term cardiac cycle refers to the sequence of events which propels blood through the heart and its associated blood vessels:

Step 1: Atrial diastole

- The atria are relaxed. The valves separating the atria from the ventricles are closed.
- The right atrium fills with deoxygenated blood from the venae cavae. The left atrium fills with oxygenated blood from the pulmonary veins.
- As the atria fill, increasing pressure is put on both valves. They start to open.

Step 2: Atrial systole

- Nerve impulses generated in the sino-atrial node (SAN) located in the right atrium spread out through the muscles of the atria.
- The atria contract. Their volume decreases so the pressure of blood inside increases.
- Blood is forced through the tricuspid and bicuspid valves into the ventricles.
- The impulses from the SAN stimulate the atrioventricular node (AVN) in the septum separating the atria.
- Impulses from the AVN pass along the bundle of His which is made up of strands of Purkyne tissue (tissue that conducts nerve impulses).

Step 3: Ventricular systole

- The wave of impulses from the bundle of His stimulates the muscle of the walls of the ventricles at their apex.
- The ventricles contract from the apex upwards. Their volume decreases so the pressure of the blood inside increases.
- This forces shut the tricuspid and bicuspid valves separating the atria from the ventricles (preventing backflow into the atria), and opens the semi-lunar valves guarding the openings of the pulmonary artery and aorta.
- Blood is forced through the pulmonary artery and the aorta.
- The elasticity of the artery walls allows for the increase in the volume of blood. Back flow into the ventricles is prevented by the semi-lunar valves.

Step 4: Ventricular diastole

- Relaxation of the ventricles marks the end of the cardiac cycle.

Volume and pressure changes during the cardiac cycle

The healthy heart at rest beats on average between 60 and 80 beats per minute: this is the heart rate.

The volume of blood pumped from the heart each minute (cardiac output) depends on the heart rate and volume of blood pumped out with each beat (stroke volume).

Heart rate, stroke volume, and cardiac output measure the heart's effectiveness and fitness:

cardiac output = heart rate × stroke volume

> **Revision tip**
> A fit heart at rest has 25% more output of blood than an unfit heart. Its stroke volume is also greater and a fit heart beats more slowly.
>
> The maximal pressure in the left ventricle is greater than in the right ventricle because the wall of the left ventricle is thicker so it contracts more powerfully.

> **Summary questions**
>
> 1 Heart valves open and close. Which valves are open and which valves are closed during atrial systole? [2]
>
> 2 Describe how the sino-atrial node and the atrio-ventricular node regulate the heartbeat. [5]

7.6 Blood vessels and their functions

Specification reference: 3.3.4.1

Blood vessels

Exchanges between capillaries and tissues, and the role of lymph

The walls of capillary blood vessels are one cell thick. As a result substances easily diffuse between blood in the capillaries and the surrounding tissues. The capillaries form dense networks called capillary beds in the tissues of the body, providing a large surface area which maximises the rate of exchange of materials between the blood and tissues. The blood in capillaries supplies nearby cells with oxygen, food molecules, and other substances. It also carries away carbon dioxide and other waste substances produced by the cells' metabolism.

> **Revision tip**
> Arteries are blood vessels which carry blood at high pressure from the heart. Veins are blood vessels which carry blood at low pressure to the heart. Capillaries are blood vessels which link arteries and veins.

> **Revision tip**
> Figure 1 shows that valves inside veins ensure that blood flows in one direction only. Remember that blood pressure in veins is lower than in arteries. Valves stop blood flowing backwards.

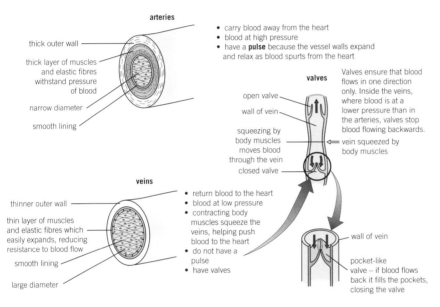

▲ Figure 1 *Valves of veins at work*

1 Hydrostatic pressure at the artery end of a capillary bed is high because of the force generated by contractions of the heart.

2 The pressure forces small molecules dissolved in the blood plasma (solutes) through the walls of the capillaries into surrounding tissues. The plasma is now called tissue fluid.

3 Water escapes through the walls of the capillaries but not large protein molecules. So the water potential of the blood is lowered (more negative). This produces a hydrostatic pressure (oncotic pressure) that opposes the hydrostatic pressure forcing solutes from the blood plasma into surrounding tissue, and has the effect of drawing solutes back into the blood capillaries.

4 At the artery end of a capillary bed, the hydrostatic pressure forcing molecules out of the capillary vessels is greater than the oncotic pressure drawing them in. As a result there is a net outflow of solute from the capillaries.

5 At the vein end of the capillary bed, the oncotic pressure is greater than the hydrostatic pressure (now reduced because of the resistance of the capillary walls to the flow of blood through the capillary vessels). As a result there is a net movement of substances in solution into the capillaries.

6 The movement of substances in solution out of the capillaries is greater than the return flow. As a result an excess of fluid bathes the tissues.

7 The excess tissue fluid drains into the lymph vessels which pass to all of the tissues of the body. The tissue fluid is now called lymph.

8 The system of lymph vessels joins the blood system. As a result the volume of the lymph in the lymph vessels remains constant.

> **Summary questions**
>
> 1 Identify some of the differences between arteries, veins, and capillary blood vessels. Explain how the differences affect function. [3]
>
> 2 Briefly describe the role of valves in veins. [2]
>
> 3 Summarise the processes which lead to the exchange of substances between capillary blood vessels and tissues. [6]

7.7 Transport of water in the xylem
7.8 Transport of organic molecules in the phloem

Specification reference: 3.3.4.2

> **Revision tip**
> The root tip, fringed with root hairs, grows through the soil. Water passes from soil into the root hairs by osmosis. The transport tissue xylem carries water and salts in solution from the roots up the stem to all parts of the plant.

Transport of water in the xylem

The passage of water across the leaf

Water moves from cell to cell by the apoplastic, symplastic, and vacuolar routes. The diffusion of water by the apoplastic route depends on the pull transmitted by the cohesive forces between water molecules. The apoplastic route accounts for most of the water moving between the cells of the leaf tissues. The evaporation of water by the symplastic and vacuolar routes depends on the difference in water potential (water potential gradient) between adjacent cells. The loss of water from the leaf depends on the difference in water potential (water potential gradient) between its air spaces and the atmosphere outside.

Overall there is a water potential gradient across the leaf from the leaf xylem to the atmosphere. The loss of water evaporation from the leaf through the stomata into the atmosphere is called transpiration.

Diffusion shells

There is a layer of stationary air called a diffusion shell adjacent to the leaf's surface. The thickness of the diffusion shell depends on the structure of the leaf. It also depends on wind speed. Water vapour diffuses across the diffusion shell before being carried away by moving air. Any factor that reduces the thickness of the diffusion shell increases the rate of evaporation of water vapour from the leaf.

The movement of water up the xylem of the stem

As water is lost from the leaf through transpiration it is replaced by more water drawn by osmosis from the xylem of the leaf into the adjacent mesophyll cells. The movement of water molecules into the tissues of the leaf draws up other water molecules through the xylem of the stem. This is because of the pull of water molecules moving from wall to wall of the cells of the leaf's tissues by the apoplastic route. The effect is called transpiration pull and produces a state of tension in the columns of water within the xylem vessels. Transpiration pull is possible because of the considerable cohesive forces between water molecules as a result of hydrogen bonding. These cohesive forces are sufficient to raise water to the tops of the tallest trees, and the theory is known as the cohesion-tension theory. In addition, adhesion is an attractive force between the water molecules and the walls of the xylem tubes. Adhesion also contributes to maintaining the transpiration stream.

> **Revision tip**
> Phloem tissue consists of sieve tubes and companion cells (sieve tubes run by the side of xylem vessels). Together, the tissues form a mass transport system.
>
> Many of the organic substances transported in solution through sieve tubes are nutrients that plants need as a source of energy and to grow, reproduce, and maintain themselves.

➕ Go further

a The table shows pressure readings during the contraction and relaxation of the aorta and different parts of the human heart.

	Pressure (kPa) in left ventricle/left atrium/aorta contracting	Pressure (kPa) in left ventricle/left atrium/aorta relaxing
Left ventricle	16.1	0.8
Left atrium	0.3	1.2
Aorta	12.2	10.6

Use the information to explain what happens in terms of pressure changes and blood flow when the:

   i bicuspid valve between the left atrium and left ventricle opens   **ii** semi-lunar valve at the base of the aorta opens.

b Why is the maximal pressure during contraction in the left ventricle greater than the maximal pressure in the right ventricle? Explain your answer.

Transport of organic molecules in the phloem

The mass flow mechanism

The following three-stage process presents a theory for the translocation of materials in the phloem.

Stage 1: Sucrose is made in the palisade tissue and tissues of the spongy mesophyll (photosynthesising tissue) of the leaf (source). Active transport transfers sucrose from the cells of photosynthesising tissues into the companion cells of the phloem tissue. From the companion cells, sucrose is actively transported into the sieve tubes nearby. The process is called loading and requires ATP. As the concentration of sucrose in the sieve tubes increases, the water potential of the solution becomes more negative. Water enters the sieve tubes by osmosis.

Stage 2: Entry of water increases the hydrostatic pressure in the sieve tubes, propelling sucrose (and other materials) through the sieve tubes. The process is an example of a pressure flow system at work.

Stage 3: In the phloem of the root the companion cells actively transport sucrose out of the sieve tubes (unloading) and into the root cells (sink) which use it as a source of energy or store it as starch. The concentration of sucrose in the sieve tubes decreases: the water potential of the solution in the sieve tubes increases water leaves the sieve tubes by osmosis the pressure in the sieve tubes decreases. The pressure difference in the sieve tubes between the source (leaves) and sink (roots) maintains the circulation of nutrients through the phloem tissue. The process is an example of mass flow.

▼ Table 1 *Mass flow: The evidence*

For	Against
Pressure difference between leaves (source) and roots (sink).	Rate of translocation of different solutes is not the same. Mass flow predicts rates should be the same.
Respiratory poisons that inhibit aerobic respiration stop translocation because insufficient ATP is produced. Loading and unloading of sucrose is prevented.	Role of sieve plates in translocation is not clear.
ATP is made in mitochondria. Companion cells have many mitochondria.	Sucrose in solution is carried to sinks at the same rate irrespective of the concentration of sucrose in the sinks. Rate of delivery of sucrose solution should be greater to sinks with low sucrose concentration compared with sinks with a higher concentration.
Increase in the concentration of sucrose solution in the sieve tubes follows soon after the increase in sucrose concentration of photosynthesising leaf cells (source).	
Concentration of sucrose solution in leaf cells (source) is greater than in root cells (sink).	
At low light intensity translocation stops because photosynthesis stops (production of sucrose stops).	

Mass transport

Key term

Translocation: The transport of nutrients (organic molecules) in solution by phloem tissue.

Key term

Sap: The nutrient solution transported through phloem tissue.

Summary questions

1. The terms 'apoplastic route' and 'symplastic route' refer to the passage of water across the roots and leaves of a dicotyledonous plant. Explain the difference between the two terms. [2]

2. Summarise the processes by which water passes from the soil, into a plant, through the plant, and into the atmosphere. [4]

3. Distinguish between transpiration and translocation. [2]

4. The process of translocation occurs because of sources and sinks. Explain how. [2]

5. Explain the role of osmosis in translocation. [4]

7.9 Investigating transport in plants

Specification reference: 3.3.4.2

> **Revision tip**
> Phloem tissue runs through the bark of woody stems.

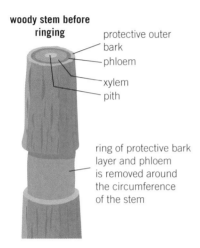

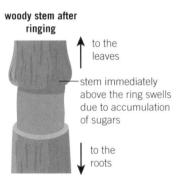

▲ **Figure 1** *Ringing of a woody stem and its results*

> **Revision tip**
> Sucrose is an important product of the reactions of photosynthesis.

Investigating transport in plants

It was understood several hundred years ago that the xylem of plants transports water. Our understanding that phloem transports nutrients is more recent. Tree ringing experiments and experiments using radioactive tracers have provided important evidence.

Tree ringing

- Ringing horizontally cuts away a strip of bark just thick enough to break the connection between the phloem of the lower part of the stem and its upper part.
- The xylem remains intact.
- Later (a few days) a bulge in the bark above the cut of the ring indicates translocation of sugars to that part of the stem and their accumulation. The absence of a bulge below the cut of the ring indicates that translocation of sugar across the cut has not taken place. Figure 1 illustrates the results of the experiment.

Aphids and radioactive tracers

Aphids (greenfly and blackfly) are small insects that feed on sap. In order to feed, an aphid is able to selectively locate the sieve tubes of phloem tissue into which it inserts its sharply pointed stylets (tube-like mouthparts). The pressure of liquid in the sieve tubes force-feeds the aphid to the point where drops of sap ooze from its anus. The set up can be used to study the translocation of sugars in phloem tissue.

- An aphid feeding on a plant is removed from its stylets. Sap continues to ooze from the stylets for several days.
- Sucrose labelled with the radioactive isotope ^{14}C is injected into a leaf of the plant.
- Samples of sap are collected from the stylets at regular intervals.
- The radioactive sucrose is a tracer. Radioactivity is detected in the samples soon after injection of the radioactively labelled sucrose.

The results indicate that nutrients (including sucrose) are transported in phloem tissue because aphids are able to selectively locate sieve tubes into which they insert their sharply pointed stylets.

> **Summary questions**
>
> 1. Aphids (greenfly and blackfly) are small insects. What characteristics of aphids make them a useful tool with which to investigate translocation? [3]
>
> 2. Explain why tree ringing experiments suggest that nutrients are transported in phloem tissue and not xylem. [3]
>
> 3. Explain how the radioactive isotope ^{14}C is used to investigate translocation. [4]

Chapter 7 Practice questions

1 The diagram shows a section through the human heart viewed from the front.

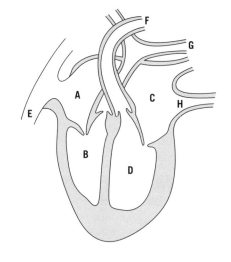

   a Name the parts of the heart and associated blood vessels labeled A–H.
   (4 marks)

   b Which *two* parts of the heart show that the chambers A and C are relaxed?
   (2 marks)

   c Why is the wall of chamber D thicker than the wall of chamber B?
   (3 marks)

2 The diagram represents the movement of tissue fluid, plasma, and lymph between capillary blood vessels, cells of tissues, and lymph vessels.

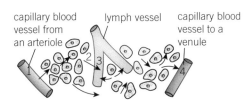

   a Name the fluid in each of the locations numbered 1–4. *(4 marks)*

   b Explain why fluid leaves the capillary blood vessels at 1, but enters the capillary blood vessel at 4. *(5 marks)*

3 The bar chart shows the rate of blood flow to different organs of the body.

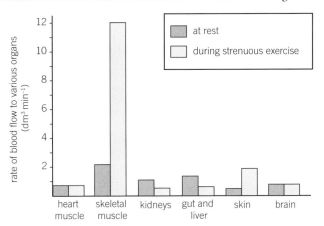

   a What is the volume of blood per minute being pumped by the left ventricle to the different organs when the body is:

   i at rest

   ii strenuously exercising? *(6 marks)*

   b Explain the link between the decrease in blood supply to the kidneys, gut, and liver and the increase in blood supply to the skeletal muscle during strenuous exercise. *(2 marks)*

4 Papers impregnated with cobalt chloride are blue when dry and pink when wet. When blue cobalt chloride papers are sandwiched against the upper and lower surfaces of some of the leaves of a plant maintained at low humidity, the cobalt chloride papers sandwiched against the lower surface of the leaves changed colour from blue to pink significantly more quickly than the papers sandwiched against the leaves' upper surface.

   a Explain why the blue cobalt chloride papers sandwiched against the lower surface of leaves changed to pink more quickly than the papers sandwiched against the upper surface of leaves. *(2 marks)*

   b At night time the cobalt chloride paper remains blue. Explain why. *(2 marks)*

8.1 Genes and the genetic code

Specification reference 3.4.1

> **Synoptic link**
> Revise the structure of a polypeptide in Topic 1.6, Proteins, and the structure and replication of DNA in Topic 2.1, Structure of RNA and DNA.

> **Key term**
> **Structural genes:** Affecting the synthesis of enzymes and the other polypeptides that make up body structures. For example collagen accounts for up to 25% of total body protein (tendons, ligaments, connective tissue, etc.).

> **Key term**
> **Regulatory genes:** Affecting the synthesis of polypeptides which control the development of organism. The polypeptides are called transcription factors (TFs). Regulatory genes may also affect the activity of other genes.

> **Revision tip**
> The sequence of bases of all of the genes of a cell, and the information each sequence carries, is the genetic code. The code is universal. It is the same in the cells of all living things.

Genes

Sequences of bases in *some* sections of DNA carry information which enables cells to synthesise molecules of polypeptide. These sections of DNA are called genes. The term locus refers to the position of a gene on a particular strand of DNA. Differences between genes are the result of differences in the sequence of their bases. The term polypeptide will be used to include peptides or proteins.

The genetic code

To make a molecule of a particular polypeptide, many amino acid units must combine in the correct order. The sequence of bases of a gene is responsible for getting the order correct.

The information needed to assemble one amino acid unit in its correct place in a polypeptide is contained in a sequence of three bases. The sequence is called a codon. We say that the genetic code is a triplet code.

A gene therefore is a sequence of codons and the genetic code is all of the codons in the DNA of a cell.

The code is also:

- **Non-overlapping:** the base of a triplet specifying the position of a particular amino acid unit does not contribute to specifying the positions of other amino acid units.
- **Degenerate:** nearly all amino acids in a polypeptide molecule are specified by more than one codon. Of the 64 possible codons:
  - three are stop codons: during protein synthesis 'stop' means 'end of polypeptide chain'
  - one codon is an 'initiator': during protein synthesis this means 'start of polypeptide chain'.

Coding and non-coding DNA

In eukaryotes most genes are not a continuous sequence of codons. Non-coding regions (called introns) split up the coding regions (called exons). Also there are non-coding regions of DNA between genes. These non-coding regions often consist of short sequences of bases which repeat over and over again (multiple repeats) and are called mini-satellites.

Summary questions
1 What does it mean when we say that the genetic code is non-overlapping? [1]
2 Distinguish between exons and introns. [2]
3 Define the term mini-satellite. [1]
4 What is a code? [1]

8.2 DNA and chromosomes
Specification reference 3.4.1

Chromosomes
The bar-shaped structures seen in an optical microscope in the nucleus of a eukaryotic cell when the nucleus divides are its chromosomes. Each chromosome consists of:

- a length of double-stranded DNA which carries many genes
- proteins called histones.

DNA and histones bind together.

The visible structure of chromosomes
In the early stages of mitosis and meiosis, strands of chromatin condense and become more compact, forming individual chromosomes. A chromosome is usually seen as a four-arm structure, consisting of a pair of chromatids joined by a centromere.

DNA in prokaryotic and eukaryotic cells
The term 'chromosome' not only refers to the four-arm structures visible during division of the eukaryotic nucleus. It also refers to the looped lengths of DNA in the cytoplasm of prokaryotic (bacterial) cells. Each looped length is called a nucleoid.

Nucleoids are not associated with histones and therefore, in strict terms, are not chromosomes even though the term is often used to describe them. Small looped molecules of DNA called plasmids are also found in the cytoplasm of prokaryotic cells.

Homologous chromosomes and alleles
A zygote (fertilised egg) inherits *two* sets of chromosomes: one set from the male parent, the other set from the female parent. Each chromosome of a set is the pair of the corresponding chromosome of the other set. The term homologous refers to the pair.

Since each chromosome is one of a pair with its homologous partner, then the genes each chromosome carries each have an homologous partner, each located at the same locus on their respective homologous chromosomes. The term allele refers to each of the homologous genes, and the term gene strictly refers to both of the alleles of the gene in question.

> **Key term**
>
> **Chromatid:** One of the two identical parts of the chromosome formed after DNA replication during the S-phase of interphase.

> **Key term**
>
> **Centromere:** The point where the two chromatids touch and which attaches to the microtubules of the spindle fibres formed during prophase and metaphase of mitosis and meiosis.

> **Revision tip**
>
> Plasmids are also found in the mitochondria and chloroplasts of eukaryotic cells.

> **Summary questions**
>
> 1. What does each chromosome of a eukaryotic cell consist of? [1]
> 2. Briefly explain the relationship between a chromosome and a pair of chromatids. [2]
> 3. What is a plasmid? [1]
> 4. Describe the process that enables a length of DNA to fit into the nucleus of a cell. [3]

8.3 Structures of ribonucleic acid
Specific reference: 3.4.1

> **Synoptic link**
> Topic 8.1, Genes and the genetic code, will help you to remember that the genetic code is not only universal, but non-overlapping and degenerate as well.

The genetic code

To make a molecule of a particular polypeptide, many amino acid units must combine in the correct order. The sequence of bases of a gene is responsible for getting the order correct. In other words, the sequence carries the information which enables a cell to assemble that particular polypeptide. The sequence of bases of all of the genes of a cell, and the information each sequence carries, is the genetic code. The code is universal. It is the same in the cells of all living things.

The information needed to assemble one amino acid unit in its correct place in a polypeptide is contained in a sequence of three bases. The sequence is called a codon. We say that the genetic code is a triplet code.

RNA structure

Like DNA, RNA is also a polynucleotide. It differs from DNA in that it is usually single stranded rather than double stranded.

- Sugar units in the chain are not deoxyribose as in DNA, but ribose (carbon number 2 has an –OH group).
- Base thymine is replaced by the base uracil. So the bases in RNA are adenine (A), uracil (U), cytosine (C), and guanine (G).

Types of RNA

Messenger RNA (mRNA) is synthesised in the cell nucleus. Its molecules are each usually single stranded each consisting of many nucleotides joined together by condensation reactions. The sequence of nucleotides (and therefore bases) forming a strand of RNA is a complement of the sequence of nucleotides (and therefore bases) of the strand of the section of DNA which is the template against which the RNA strand forms.

Transfer RNA (tRNA): tRNAs are smaller molecules than mRNA. The shape is held by the hydrogen bonding of complementary bases within the molecule.

- An amino acid is attached at one end of the molecule which ends in the base sequence CCA. This sequence forms the amino acid accepting site. Transfer RNAs act as carriers of amino acids during polypeptide synthesis. Each type of amino acid is carried by its own type of tRNA.
- There are at least 20 different amino acids. There are, therefore, at least 20 different types of tRNA.
- Three bases at the other end of the tRNA molecule (the anticodon) are complementary to the mRNA codon encoding the amino acid it carries.

Ribosomal RNA (rRNA) combines with protein forming ribosomes. Each ribosome is a site where mRNA and tRNA interact, synthesising polypeptides.

> **Revision tip**
> The term genome refers to all of the DNA and its sequence of bases in the cells of an organism. The human genome consists of 3.1 billion (10^9) bases. Less than 3% are base sequences (genes) encoding expression of proteins. The other base sequences either regulate gene expression, are repeat sequences, or are currently without known function.

> **Synoptic link**
> Topic 8.4, Polypeptide synthesis: transcription and splicing, and Topic 8.5, Polypeptide synthesis: translation will help to explain the link between RNA and the genetic code.

Summary questions	
1 Compare the structure of mRNA and tRNA.	[2]
2 Distinguish between the roles of mRNA and tRNA in polypeptide synthesis.	[2]

8.4 Polypeptide synthesis: transcription and splicing

Specification reference 3.4.2

Polypeptide synthesis is a two-stage process. The first stage is **transcription**. That is, the synthesis of mRNA from the DNA template of a gene. The information carried in the sequence of bases of the gene is copied in the complementary sequence of bases of the mRNA transcribed.

Transcription: the process

Step 1

The enzyme DNA helicase catalyses the breaking of the hydrogen bonds that link the base pairs of a double-stranded section of DNA carrying the gene to be transcribed. The DNA unwinds and unzips, and its strands separate. The bases of each strand are exposed.

Some base sequences of one of the unzipped strands (the transcribing strand) carry the genetic information which encodes polypeptide synthesis. Its unzipped partner strand does not carry genetic information.

Each sequence of bases of the transcribing strand which encodes the synthesis of polypeptide is called an exon.

The sequences of bases that do *not* encode polypeptides are called introns (bacterial DNA does not contain introns).

Step 2

mRNA is synthesised on the transcribing DNA strand. The enzyme RNA polymerase first binds to a promoter sequence on the transcribing strand, initiating transcription. The enzyme then moves along the transcribing strand, adding RNA nucleotides to the growing mRNA strand. The bases of the nucleotides are complementary to the exposed bases of the transcribing strand.

A cap of G (guanine) is added to the 5' end, and a poly-A tail (150–200 adenines) to the 3' end of the pre-mRNA molecule.

The growing mRNA strand lengthens in the 5'→ 3' direction.

A strand of pre-mRNA (precursor mRNA) is produced. Pre-mRNA carries exons and introns which are complements of the exons and introns of the transcribing strand.

In eukaryotic cells, further processing occurs which adds signalling sequences to, and removes introns from, the pre-mRNA. The process is called editing. Enzymes called spliceosomes catalyse the splicing (joining together of) the exons. Lengths of RNA called ribozymes have the same effect. In other words pre-mRNA can catalyse its own editing. The result is a strand of mature mRNA. The mRNA of prokaryotic cells (bacteria) does not carry introns. Transcription is direct and editing does not occur.

Step 3

Strands of mature mRNA pass from the nucleus through the nuclear pores of the nuclear membrane into the cytoplasm of the cell. Its poly-A tail facilitates (makes easy) the passage of each strand.

> **Synoptic link**
>
> For more information on the second stage of the polypeptide synthesis process see Topic 8.5, Polypeptide synthesis: translation.

> **Summary questions**
>
> 1 Explain the roles of DNA helicase and RNA polymerases in transcription. [2]
>
> 2 Explain how pre-mRNA can catalyse its own editing. [2]

8.5 Polypeptide synthesis: translation
Specification reference 3.4.2

> **Synoptic link**
>
> For more information on the first stage of the polypeptide synthesis process see Topic 8.4, Polypeptide synthesis: transcription and splicing.

Polypeptide synthesis is a two-stage process. The second stage is **translation**. That is the conversion of the information in mRNA to make a polypeptide. The information is carried in the sequence of bases of the mRNA (itself a complementary copy of the base sequence of the gene against which the mRNA was transcribed) and determines the sequence in which amino acids join together forming a polypeptide.

Translation: the process

Step 1

A strand of mature mRNA binds to a ribosome. Binding is facilitated by its poly A-tail. Its 5' cap signals the point of attachment. The ribosome moves along the strand until it reaches an initiation codon. This is usually AUG and signals the beginning of a gene.

Step 2

Molecules of tRNA collect amino acids from the 'pool' of amino acids dissolved in the cytoplasm of the cell; each type of amino acid is carried by its own type of tRNA. The combination of an amino acid with its particular tRNA requires energy released by the hydrolysis of ATP. The process is called activation. The tRNA/amino acid combinations move towards a ribosome.

Step 3

The tRNA /amino acid combination which carries the anticodon UAC combines with its complement, the first codon AUG. This is initiation and translation begins. The second codon of the mRNA then attracts its complementary tRNA anticodon in a second tRNA/amino acid combination. The ribosome holds the two combinations in place while a peptide bond forms between the two amino acids. The reaction is catalysed by the enzyme peptidyltransferase.

Once the peptide bond forms, the bond between the first (initiation) molecule of tRNA and its amino acid is hydrolysed and the unbound tRNA is released from its complementary mRNA codon. The third codon of the mRNA attracts its complementary tRNA anticodon, in a third tRNA/amino acid combination. The ribosome moves along to hold it in place while a peptide bond forms between the amino acid which its tRNA carries and the second amino acid. The bond between the second molecule of tRNA and its amino acid is hydrolysed and the unbound tRNA is released from its complementary mRNA codon.

As a result a strand of polypeptide forms, one amino acid at a time, according to the particular sequence of codons of the mRNA coding tRNA/amino acid combinations to assemble in a particular order. The process is called **elongation**.

Step 4

The ribosome moves along the length of mRNA until it reaches a stop codon: UAA, UGA, or UAG. This codon does not attract an anticodon but encodes a releasing factor. The bond between the terminal (end) tRNA molecule and its amino acid is hydrolysed and the completed polypeptide molecule released. The process is called termination.

DNA, genes, and protein synthesis

Polypeptide chains may then fold to form secondary and tertiary structures. Several chains may combine to form a quaternary structure. Polypeptides may also undergo post-translational modification. For example, the initiator codon AUG encodes the amino acid methionine. Not all polypeptides begin with methionine. Post-translational modification removes methionine at the beginning of the polypeptide chain.

Polypeptides used inside cells are usually made on free ribosomes and released into the cytoplasm.

Step 5

Polypeptides to be exported from cells are usually made on the ribosomes of the rough endoplasmic reticulum and transported to the Golgi apparatus. Here they are modified and packaged into vesicles which bud off from the Golgi. The vesicles pass to the plasma membrane from which they are secreted by exocytosis or where they form membrane proteins.

siRNA interferes with gene activity

Small interfering (si)RNA is one of several types of RNA molecule that help to regulate which genes are active and how active they are. Each molecule consists of a short sequence of RNA about 20 nucleotides long.

- The binding of siRNA with its complementary length of mRNA causes the mRNA to break.
- As a result translation is prevented and the gene is silenced.

Artificial siRNAs are tailor-made to be complementary to the mRNA of different genes. This means that siRNA can be used to silence specific genes. For example, silencing an oncogene would switch off the overexpression of the polypeptide stimulating mitosis. In theory the rate of mitosis would slow, perhaps stopping the development of a tumour. Our ability to selectively silence genes promises exciting developments in medicine and other areas of research.

> **Revision tip**
> Ribosomes formed from rRNA and protein are found free in the cytoplasm or bound to the endoplasmic reticulum. Each consists of a small subunit which binds mRNA and a large subunit which binds tRNA. The combination of a strand of mRNA with a number of ribosomes (5–50) forming a polysome speeds up polypeptide synthesis.

Summary questions

1. Distinguish between transcription and translation. [2]
2. What is a polysome? [1]
3. Describe the role of ATP during translation. [2]

Chapter 8 Practice questions

1 The letters represent the sequence of bases of a short length of a strand of DNA. The letters do not overlap.

AATCCTGACTAGGAT

    a Explain the significance of the statement "The letters do not overlap."

(2 marks)

    b Explain why the substitution of one of the bases in the strand with another different base need not necessarily alter the sequence of amino acids. *(2 marks)*

    c If the number of amino acid units form a whole molecule, which term would best describe the molecule: protein, peptide, or polypeptide?

(1 mark)

2 The figure shows a short sequence of bases of DNA. The sequence encodes the position of a number of amino acids in a molecule of polypeptide.

AATGCACTATACCCGGCC

    a i How many amino acids are encoded by the sequence of bases?

(1 mark)

        ii Explain your answer to part **i**. *(2 marks)*

    b Write the sequence of bases of the part of mRNA transcribed against the DNA sequence in the figure. *(1 mark)*

    c Describe the relationship between your written sequence of mRNA in part **b** and the DNA sequence in the figure. *(2 marks)*

    d Write the sequence of the bases of the anticodons of the tRNA molecules which bind to the sequence you have written for the mRNA in part **b**. *(1 mark)*

    e Describe the relationship between your written sequence of anticodons of tRNA molecules in part **d** and the DNA sequence in the figure. *(2 marks)*

    f Explain why the number of bases in a sequence of mature mRNA may be fewer than the number of bases in the sequence of DNA against which the mRNA was transcribed. *(3 marks)*

3 Describe the relationship between chromosomes, chromatids, DNA, and histones. *(5 marks)*

9.1 Gene mutation

Specification reference: 3.4.3

Gene mutation

Gene mutations are the result of copying errors in the sequence of bases of their DNA. They occur during DNA replication. If only one base is involved, then the copying error is called a point mutation. There are two types:

- base pair deletions (bases are lost) or insertions (bases are added)
- base pair substitutions (bases are replaced with others).

Deletions or insertions alter the sequence of the bases downstream of the locus (position of a gene on a chromosome) on the DNA strands where the mutation takes place. We say that the mutation causes a frameshift, and frameshift mutations can significantly affect the structure and therefore function of polypeptides.

Substitutions also alter the sequence of bases within a codon, but do not do so downstream of the mutation. The alteration may change the amino acid unit at the point of the mutation, but it may not. The genetic code is degenerate. This means that if the substitution replaces a base within a codon with a base of an alternative codon encoding the same amino acid, then the amino acid sequence of the polypeptide as a whole will not change, even though the base sequence of the codon has. This sort of mutation is called a silent mutation.

Chromosome mutations

Chromosome mutations occur when chromosomes break in the early stages of meiosis. Breakage may cause:

- Deletion: a part of a chromosome is lost.
- Translocation: the broken part of a chromosome joins to another chromosome.
- Inversion: the broken part of a chromosome rejoins the chromosome but after turning through 180°.

One or more whole chromosomes may be lost. For example, during meiosis **non-disjunction** produces two sorts of sex cell: one sort does not have any copies of the chromosome in question, the other sort has two.

Where non-disjunction of chromosome number 21 occurs during meiosis in one parent, the result is referred to as Down syndrome. The affected individual has an extra copy of chromosome 21 in each body cell.

> **Synoptic link**
>
> First revise DNA structure and replication (Topic 2.1, Structure of RNA and DNA, and Topic 2.2, DNA replication), genes and the triplet code (Topic 8.1, Genes and the genetic code), and chromosome structure (Topic 8.2, DNA and chromosomes).

> **Revision tip**
>
> The term **mutation** refers to a change in the arrangement or the amount of genetic material in a cell. Mutations in sex cells are inherited. Mutations in other (body) cells are *not*. We say that the mutations are acquired.

> **Revision tip**
>
> Something that causes gene mutation is called a mutagen. If the effect of a **mutagen** results in an oncogene (a mutated gene that leads to cancer) then the mutagen is called a **carcinogen**. Examples of carcinogens include different substances in cigarette smoke and ionising radiation.

> **Revision tip**
>
> The background mutation rate is the natural rate of spontaneous mutation of a particular gene.

Summary questions
1 What is a point mutation? [1]
2 Explain why a mutation might be silent. [2]
3 Explain how a frameshift alters the amino acid sequence of the polypeptide encoded by the mutated gene downstream of the mutation. [2]

9.2 Meiosis and genetic variation
Specification reference: 3.4.3

> **Synoptic link**
> The outcome of mitosis was covered in Topic 3.7, Mitosis. Chromosomes were covered in Topic 8.2, DNA and chromosomes.

Meiosis and generic variation

Cell division by meiosis gives rise to gametes (sex cells). Gametes are haploid rather than diploid; an important difference between meiosis and mitosis. Also, unlike cells produced by mitosis, the gametes produced by meiosis are *not* genetically identical with each other and their parent cell. Each is genetically unique. In meiosis, the process passes through the same phases as mitosis (with some differences), but the phases occur twice over.

- The first meiotic division is a reduction division which results in two daughter cells, each with half the number of chromosomes of the nucleus of the parent cell. The cells are haploid.
- The second meiotic division is a mitosis during which the two haploid daughter cells (from the first meiotic division) divide, resulting in four haploid daughter cells (gametes).

Only the cells that give rise to gametes (the sex cells) divide by meiosis. Sex cells are produced in the sex organs:

- the testes of the male and ovaries of the female in mammals
- the anthers (male) and the carpels (female) in flowering plants.

> **Revision tip**
> During fertilisation (when a sperm joins with an egg), the chromosomes of sperm and egg combine. Fertilisation is a random process: any sperm can combine with any egg.

> **Revision tip**
> Remember that during cell division the chromosomes in the nucleus of the parent cell pass to the new daughter cells.

> **Revision tip**
> During meiosis chromosomes separate randomly. As a result alleles (genes) separate randomly. The process is called independent segregation/assortment.

How much genetic variation is generated from meiosis and fertilisation?

The possible number of combinations of chromosomes in any gamete, depending on the number of chromosomes in the parent cell, can be calculated using the formula: $2n$

where n = the number of paired homologous chromosomes.

> **Revision tip**
> Adjacent chromatids of a pair of homologous chromosomes (a bivalent) break and exchange pieces of the chromatids in question. The process is called **crossing over** and results in new combinations of genes on the chromatids.

> **Worked example**
>
> In Figure 1 each cell carries two pairs of homologous chromosomes. How many different combinations of the chromosomes are possible?
>
> The possible number of different combinations of homologous chromosomes is:
>
> 2^2 or 4 possible different combinations of chromosomes (each combination consisting of one paternal and one maternal chromosome).
>
> Most human cells carry 23 pairs of homologous chromosomes.
>
> Therefore the possible number of different combinations of homologous chromosomes is:
>
> 2^{23} or 8,388,608 possible different combinations of chromosomes (each combination consisting of one paternal and one maternal chromosome).

Genetic diversity and adaption

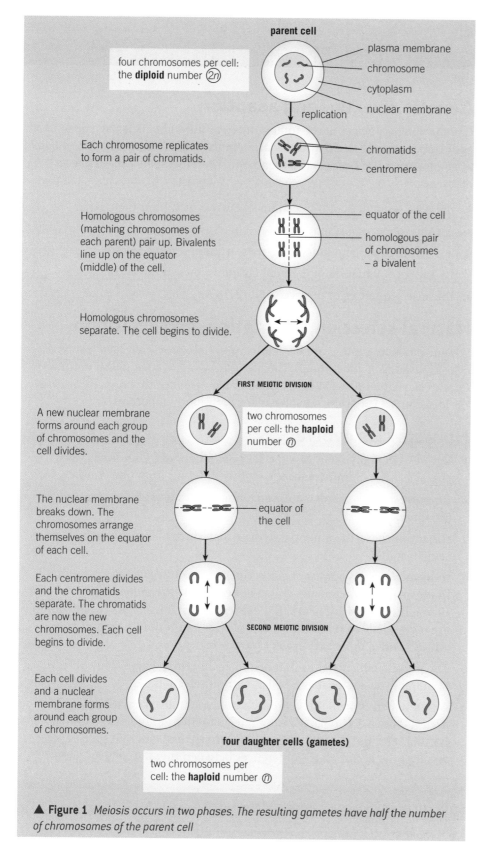

▲ **Figure 1** *Meiosis occurs in two phases. The resulting gametes have half the number of chromosomes of the parent cell*

Random pairing of sperm and egg at fertilisation increases the possible chromosome combinations in offspring. The possible number of these combinations can be calculated using the formula $(2n)^2$

where n = the number of paired homologous chromosomes.

Summary questions

1. Explain the importance of meiosis. [2]

2. What is the result of meiosis? [1]

3. Why are the genetic causes of variation inherited? [2]

9.3 Genetic diversity and adaptation

Specification reference: 3.4.4

> **Revision tip**
> A population is made up of a group of individuals of the same species living in the same place at the same time. The total of all the alleles of all the genes of a population is called a gene pool.

> **Revision tip**
> A population is a pool of genes. Gene flow within a gene pool refers to the transfer of parental genes to offspring when parents reproduce sexually.

Genetic diversity and adaption

Genetic diversity refers to the total number of different alleles in a gene pool, and is represented by the variation in the genetic material of members of a population. Differences between population members can be due to differences in:

- the base sequences of the sections of DNA which form their genes
- chromosomes as a result of:
  - a crossing over and independent assortment during meiosis
  - b the recombination of parental chromosomes in the zygote
- the base sequences of the non-coding DNA.

Natural selection, adaptation, and evolution

Present-day living things are descended from ancestors that have evolved over thousands of generations. The lifetime's work of the English naturalist Charles Darwin (1809–1882) provided much evidence that organisms evolve. His ideas on *how* organisms evolve were even more important.

How species evolve

1. Because the individuals of a species population vary genetically, the characteristics of individuals are slightly different from one another.
2. Resources (food, space, mates) enabling species populations to survive are finite.
3. Organisms have the potential to over-reproduce but on average numbers remain stable.
4. Individuals with variants of genes (including advantageous mutations) that express characteristics which better enable the individuals to survive than individuals with genes that express the characteristics less favourably are more likely to reproduce. Their offspring inherit the variant genes which control those favourable characteristics.
5. The process which results in individuals inheriting genes that control favourable characteristics is called natural selection.
6. Genes which control the expression of favourable characteristics accumulate in a population from one generation to the next through natural selection. Eventually populations become different species as a result of the accumulated differences between them, that is speciation.

Summary questions	
1 What does the phrase 'genetic diversity' mean?	[2]
2 What is the difference between gene flow and gene pool?	[2]

9.4 Types of selection
Specification reference: 3.4.4

Types of selection

Natural selection works on genetic variation in the gene pool. Alleles which favour survival are selected for and their frequency changes. The characteristic(s) controlled by the favourable alleles change(s) enabling individuals with the alleles to adapt to a changing environment. The population evolves. The extent of change in the frequency of an allele (or alleles) is a measure of the rate of evolution of the population in question. Change in an environment stimulating evolution represents selection pressure.

Trends in natural selection

In stable environments selection pressure is low. For any particular characteristic (e.g. length of body), individuals with extremes of the characteristic (bodies which are very short or long) are selected against. Those close to the average are selected for and are therefore more likely to reproduce (differential reproductive success), passing on the alleles controlling 'averageness' to the next generation. If the trend is long term, the species changes very little. Descendants, therefore, look like their distant ancestors. This kind of selection maintains the constancy of a species. It is called stabilising selection.

Where the environment is rapidly changing, selection pressure is high and new species quickly arise. Selection pressure promotes the adaptation of individuals to the altering circumstances.

For example, a longer body may favour survival if predators find that catching shorter-bodied individuals is relatively easy. In these circumstances individuals with short bodies are selected against and individuals with long bodies are selected for. Long-bodied individuals are therefore more likely to reproduce (differential reproductive success), passing on the alleles for long body to the next generation. If the trend is long term, the species changes. Descendants therefore do not look like their ancestors and become a distinct and different species. This kind of selection is called directional selection. Once the new characteristic is at its optimum (maximises the survival chances of the individuals) with respect to the new environment, then stabilising selection takes over.

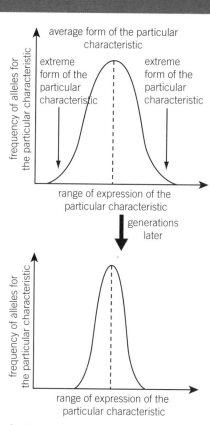

▲ **Figure 1** Over time, stabilising selection reduces the frequency of the alleles for the extreme expression of a particular characteristic (e.g. length of body)

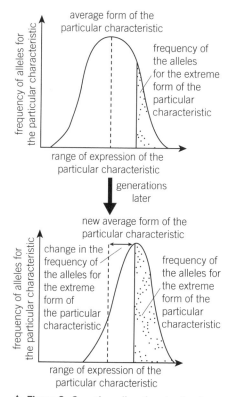

▲ **Figure 2** Over time directional selection increases the frequency of the alleles for an extreme expression of a characteristic (or characteristics) which become(s) the "new average" of the evolving species

Question and model answer

Q Evolution is sometimes defined as a result of changes in the frequency of particular alleles. Explain the definition in terms of natural selection.

A *In the case of stabilising selection, the frequency of alleles controlling the 'average' form of the characteristic(s) in question increases, whereas the frequency of the alleles controlling the 'extreme' forms of the characteristic decreases.*
In the case of directional selection, the frequency of the alleles controlling one of the extreme forms of the characteristic increases but the frequency of alleles controlling the 'average' form and the other 'extreme' form of the characteristic(s) decreases.

Summary questions

1. Explain the different evolutionary outcomes of stabilising selection and directional selection. [3]
2. Define evolution. [1]
3. What is natural selection? [2]

Chapter 9 Practice questions

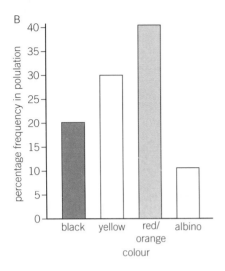

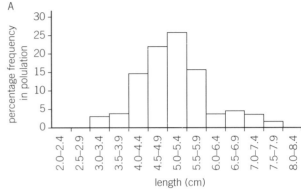

1 In a population of 300 goldfish, variations in two characteristics were measured and the results represented as charts. Chart A shows the variation in length of fish and chart B shows variation in their colour.

   a Which chart shows:

   i continuous variation

   ii discontinuous variation?

   Briefly give reasons for your answers (*4 marks*)

   b Use chart B to calculate the percentage of yellow goldfish in the population. (*1 mark*)

   c Albino goldfish are relatively rare compared with the colours of the other fish. Give a genetic explanation for the occurrence of albino goldfish. (*1 mark*)

Rose bush	Length of fruit (mm)
1	14, 18, 13, 16, 16, 13, 17, 19, 11, 19
2	12, 15, 19, 18, 15, 16, 18, 14, 19, 19
3	16, 15, 16, 14, 18, 14, 15, 18, 20, 15
4	15, 17, 14, 12, 16, 20, 19, 17, 15, 16
5	15, 13, 15, 19, 18, 20, 14, 19, 19, 14

2 Rose flowers form fruits following fertilisation. All the fruits from five rose bushes of the same species were collected and their lengths measured. The results are shown in the table:

   a Draw a graph or chart to represent the data. Explain your choice. (*3 marks*)

   b Variation in the length of the rose fruits is due to environmental factors. Describe the possible causes for the variation in fruit length. (*2 marks*)

   c Some of the seeds inside the rose fruits germinate and grow into rose bushes, which also produce fruits. Explain why the variation in fruit length of the parent rosebush due to environmental factors will not affect the fruit length of their offspring. (*4 marks*)

3 Phenylthiocarbamide is a substance which tastes bitter to some people but not others. The table shows the frequency of 'non-tasters' in different groups of people.

Group	'Non tasters' (%)
Hindus	33.7
Danish	32.7
British	31.5
Spanish	25.6
Portuguese	24.0
Japanese	7.1
Lapps	6.4
West Africans	2.7
Chinese	2.0
S. American (indigenous people)	1.2

   a Is people's response to phenylthiocarbamide an example of continuous or discontinuous variation? (*1 mark*)

   b Give a genetic explanation for the differences in responses to phenylthiocarbamide between different groups of people. (*3 marks*)

10.1 Species and taxonomy
Specification reference: 3.4.5

Classification

Organisms which have characteristics in common are grouped together. Placing organisms into groups is called classification. Some characteristics are unique to a group; there is no overlap with other groups. Other characteristics are shared with other groups. Groups therefore combine to form larger groups forming a hierarchy of taxonomic ranks. In the five kingdom system, the highest taxonomic rank is the kingdom. The kingdoms are: protists (single-celled organisms), animals, plants, fungi, and bacteria.

In descending order of size of category each: kingdom includes a number of phyla (plural); phylum (singular) includes a number of classes; class includes a number of orders; order includes a number of families; family includes a number of genera (plural); genus (singular) includes one or more species.

Taxonomy

The term taxonomy refers to the strict methods and rules of classification. For example the genus and the species identify the individual living thing with a two-part name, for example humans. Humans belong to the genus *Homo* and have the species name *sapiens*. The genus name begins with a capital letter and the species name with a small letter.

What is a species?

The usual reply would be that all the members of a population belong to the same species. The individuals are very similar to each other and can sexually reproduce offspring, which are themselves able to reproduce.

Ultimately the difference between species arises from the differences between their genes. But how great must the difference be before individuals are no longer varieties of the same species but different species? The problem of definitions means that there is no completely satisfactory answer to the question 'What is a species?'

Courtship behaviour

Courting behaviour refers to the interactions between males and females of the same species which lead to copulation, the conception of offspring, and in many species their subsequent care. Its components include:

- attraction of mates
- behaviour which enables individuals to recognise a potential partner as the opposite sex and correct species
- synchronisation of sexual behaviour between partners
- care of the offspring by one or both parents.

One component leads to the next; the behaviour of an individual causes a response in the partner. For example, the colourful feathers of male birds are often a sign stimulus which causes: **a** initially aggressive behaviour between males **b** attraction between males and females.

A sign stimulus in one individual releases behaviour in another individual. The behaviour is also a sign stimulus to which the first individual responds and so on.

> **Key term**
>
> **Phylogeny:** The relationship between organisms as a result of their shared characteristics which link them to a common ancestor.

> **Revision tip**
>
> There are alternative systems of classification. In one system the domain is the highest taxonomic rank. There are three domains: Bacteria, Archaea, and Eukarya.

> **Revision tip**
>
> If the sign stimuli and the response to them are inappropriate, then the sequence of courtship behaviour breaks down. Courtship stops and sexual activity does not take place nor therefore conception.

Biodiversity

If sign stimuli and the responses to them are inappropriate, then the sequence of behaviour breaks down. Courtship stops and sexual activity does not take place nor therefore conception. Courtship behaviour:

- enables individuals of the species to recognise one another
- promotes pre-zygotic isolation (i.e. the chances of sexual activity between individuals of different species leading to conception is reduced); as a result the chances of hybridisation are reduced and the characteristics which are unique to and identify a species are maintained.

However, there are many exceptions and the meaning of the term 'species' is problematical.

Comparing variations in haemoglobin using immunological techniques

- B-cells in the blood produce antibodies in response to the presence of antigens. This is the immune response.
- Antigens are substances to which antibodies bind. Proteins can act as antigens. Antibodies therefore can bind to proteins.
- Antibody–protein binding depends on the shape of the antigen binding sites of the antibody molecule matching the shape of the protein (or part of protein). The closer the match between antibody and protein the more strongly they bind together.
- The shape of a protein depends on the sequence of its amino acids. If one amino acid or more of a particular protein changes then the binding properties of its matching antibody will also stop. The more changes in the amino acid sequence of the protein the less strongly will the particular antigen bind to the different variants of the protein.

One way of working out the relatedness of the species to humans is to measure how strongly the antibody which is specific for the β-chain of human haemoglobin binds to the haemoglobins of the other species.

- The stronger the binding the more closely related is the species to humans.
- The weaker the binding, the more distant is the relationship.

Comparing the base sequences of DNA nucleotides

Working out the base sequences of DNA nucleotides (and the genes they form) and how the base sequences vary is one aspect of **structural genomics**. By comparing DNA base sequences among species (**comparative genomics**), it is possible to:

- identify changes in genes (and whole genomes) as species have evolved.
- work out evolutionary relationships among organisms: closely related organisms would be expected to have similar DNA base sequences; less closely related organisms would be expected to have less similar DNA base sequences.

Summary questions

1. Explain the difference between the terms classification and taxonomy. [2]
2. Why do most systems of classification reflect the phylogeny (evolutionary history) of organisms? [3]
3. What are sign stimuli? [2]
4. How does courtship behaviour promote pre-zygotic isolation? [3]

10.2 Diversity within a community
Specification reference: 3.4.6

Biodiversity and the diversity index

Biodiversity stands for 'biological diversity'. The classifications of taxonomists are inventories of biodiversity. The term refers to the diversity of:

- species
- habitats
- genes within a species population (intraspecific genetic diversity)
- genes within a group of species populations (interspecific genetic diversity).

Species richness is the number of species present in a habitat. However, this takes no account of how many individuals of each species are present. A species with a population of four counts the same as a population of hundreds.

Species evenness takes account of this drawback and it compares the number of individuals of each species. It quantifies how equal the numbers are. If they are fairly even, with all species having similar numbers of individuals, then the species evenness is high; and if they are very different in number as in the example above, the species evenness is low.

Simpson's diversity index (D) takes into account the relationship between all of the species in a community of organisms and the numbers of each species. The index can be calculated from the formula

$$D = \frac{N(N-1)}{\Sigma n(n-1)}$$

Where:

D = Simpson's diversity index

Σ = sum of

N = total number of organisms of all species

n = total number of organisms of each species

- A high value of D suggests an environment which is stable in the long term (e.g. tropical rain forests, coral reefs), and highly biodiverse.
- A low value of D suggests an environment which is unstable in the long term (e.g. hot deserts, fields used to grow crops), with relatively few species.

The index is often used to assess environmental health:

- a high value of D suggests an environment in good biological health
- a low value of D suggests an environment in poor biological health.

> **Revision tip**
> **Uncertainties of definition**
> The diversity index measures both species and genetic diversity. However, the uncertainties of the relationship between genetic diversity and the definition of species make the use of the diversity index as a measure of biodiversity a problem.

> **Worked example**
> The first step to calculating Simpson's diversity index for a particular habitat is the fieldwork which collects data on the numbers of individuals of *all* species and the numbers of individuals of each species in the habitat.
>
> Using the collected data calculate:
>
> - N and $N(N-1)$
> - $\Sigma n(n-1)$

Biodiversity

The numbers of individuals of all species and numbers of individuals of each species of a water meadow are recorded in Table 1:

▼ Table 1

Plant species	Number of individuals of each species	$n(n-1)$
Lady's smock	34	34(33) = 1122
Purple moor grass	934	934(933) = 871422
Meadow fescue	234	234(233) = 54522
Yorkshire fog	733	733(732) = 536556
Cocksfoot	2003	2003(2002) = 4010006
Ragged robin	32	32(31) = 992
Total	N = 3970 $N(N-1)$ = 15756930	$\Sigma n(n-1)$ = 5474620

The totals at the foot of each column of Table 1 are then put into the equation

$$D = \frac{N(N-1)}{\Sigma n(n-1)}$$

where

$$D = \frac{15756930}{5474620}$$

$$= 2.8$$

This is the value of the diversity index for this particular water meadow. The D value for another meadow (a meadow cut for hay) was 2.1.

The difference in D value between the meadows suggests that the water meadow is the more stable habitat because it is more biodiverse than the hay meadow. Cutting hay destabilises that habitat reducing its biodiversity. The lower D value of the hay meadow is a measure of the effect of human activities on the habitat.

Go further

Natural selection is the cause of changes in the relative frequency of genes within a population from generation to generation. The changes in gene frequency increase the representation of particular genotypes (and therefore phenotypes) in subsequent generations. The changes are adaptive, affecting 'fitness' of the population.

- **a** What is meant by the phrase 'relative frequency of genes within a population'?
- **b** Explain how natural selection causes change in the gene frequencies in a population from generation to generation.
- **c** How would you define the 'fitness' of a population?
- **d** Explain when changes in the gene frequency of a population may not be the result of natural selection.

Summary questions

1 What does Simpson's diversity index measure? [2]

2 How may uncertainties about the definition of species affect the reliability of the index of diversity? [3]

10.3 Species diversity and human activities
Specification reference: 3.4.6

Human impact on species diversity

The extinction of species through human activities reduces species diversity and so reduces genetic diversity. It also reduces the possibility of exploiting genetic material for the improvement of crops and the development of new medicines. Continuing growth of the human population increases pressure for yet more land to produce the food needed to feed people. Habitats continue to be destroyed along with the species occupying them.

Deforestation

Cutting down large areas of forest prevents regeneration of the plants removed. Destruction of the plants destroys the habitats of animals and other organisms. As a result species are driven to extinction and species diversity is reduced.

Intensive farming

The ethical issues arising from modern intensive farming include:

- Costs to the environment and therefore species diversity.
- The demand for land, reducing habitat diversity.
- Selective breeding which reduces gene diversity because variation within species is reduced to the varieties which are most productive. For example meat, milk, and abundance of grain.
- Use of pesticides which are poisonous and kill other species of wildlife as well as pest species (they may also be a hazard to human health).
- The use of fertilisers which drain in solution from land into water causing eutrophication (they may also be a hazard to human health).
- The use of modern farm machinery which works most efficiently in large open fields which results in the farming landscape being cleared of hedgerows, copses, and woods.

Conservation and agriculture

Selective breeding of crops and livestock has greatly reduced the genome in domestic populations. Conserving wild species maintains the pool of genes available for future breeding programmes to develop new breeds. As the human population grows and the demand for food increases, wild species also give us the opportunity to develop breeds suited to farming in new environments that were previously wild.

Climate change

Climate change is resulting in changes in distribution of diseases. For example, global warming and increased rainfall increase the abundance and distribution of mosquitoes and therefore the spread of malaria and other diseases that mosquitoes carry. Fungal plant diseases thrive in warmer, wetter conditions, reducing harvests. Wild animal and plant species have evolved along with these diseases so we can incorporate the genes from disease-resistant wild populations to provide disease resistance in new breeding or genetic modification programmes.

> **Revision tip**
> The methods of intensive farming reduce the biodiversity of ecosystems therefore reducing genetic diversity as well.
>
> If the conditions in which crops and animals are raised alter in the future, then the selective breeding of new varieties able to survive the changed conditions may not be possible because there is a smaller pool of genes to draw from.

> **Summary questions**
>
> 1 Explain how the human use of resources in a sustainable way helps to maintain species diversity. [2]
>
> 2 Describe how conserving wild species benefits agriculture. [2]
>
> 3 Explain why conserving habitats helps to conserve species. [3]

10.4 Investigating diversity
Specification reference: 3.4.7

Comparing the base sequences of DNA nucleotides

Data used to identify homologous characteristics (characteristics of species shared as the result of common ancestry) have traditionally come from studies of observable characteristics such as body structures.

By comparing DNA base sequences among species (**comparative genomics**), it is possible to identify changes in genes (and whole genomes) as species have evolved and work out evolutionary relationships among organisms: closely related organisms would be expected to have similar DNA base sequences; less closely related organisms would be expected to have less similar (more diverse) DNA base sequences.

Because the base sequence of mRNA is a complement of the base sequence of the strand of DNA against which the mRNA is transcribed, it follows that DNA (and therefore genetic) diversity is reflected in comparing the mRNA sequences of different species.

Comparing amino acid sequences in specific proteins

Different forms of the protein haemoglobin are an example of molecular homology. In other words, the haemoglobins found in different species are a shared characteristic inherited from a common ancestor.

- Species are more closely or more distantly related (diverse) to each other depending on the number of amino acid differences between species.
- Generally the number of amino acid differences is inversely proportional to the closeness of the relatedness – that is, the fewer the differences, the more closely related are the species; the more diverse the differences, the more distantly related are the species.

Comparing variations in haemoglobin using immunological techniques

B cells in the blood produce antibodies in response to the presence of antigens. This is the immune response.

Antigens are substances to which antibodies bind. Proteins can act as antigens. Antibodies therefore can bind to proteins.

Antibody–protein binding depends on the shape of the antigen binding sites of the antibody molecule matching the shape of the protein (or part of the protein). The closer the match between antibody and protein, the more strongly they bind together.

The shape of a protein (tertiary structure) depends on the sequence of its amino acids. If one amino acid or more of a particular protein changes then the binding properties of its matching antibody will alter. The more changes in the amino acid sequence of the protein, the less strongly will the antigen bind to the different variants of the protein.

> **Revision tip**
> One way of sequencing the bases of DNA is the chain-termination method. The process begins by breaking up the chromosomes carrying the DNA into pieces. Sequencing of the DNA bases then begins.

> **Summary questions**
>
> 1 Explain why comparing DNA base sequences is the preferred method of establishing the phylogenetic relationships between organisms. [2]
>
> 2 Explain why genetic diversity can be investigated by comparing the base sequences of mRNA as well as the DNA sequences of different species. [2]
>
> 3 Explain why the greater the variation in the sequence of bases of the DNA of different species, the more distantly related are the species. [3]

10.5 Quantitative investigations of variation

Specification reference: 3.4.7

Continuous variation

Some characteristics show variations spread over a range of measurements. Height is an example. All intermediate heights are possible between one extreme (shortness) and the other (tallness). We say that the characteristic shows continuous variation.

Discontinuous variation

Other characteristics do not show a spread of variation. There are no intermediate forms but distinct categories. For example most humans belong to blood groups A, B, AB, or O. We say that the characteristics show discontinuous variation.

Discontinuous data is represented as a bar chart because there are no intermediate values between the bars to indicate trends in the data. We say that the values indicate one circumstance or an alternative. There is no 'in-between' circumstance.

What is standard deviation?

The distribution curve of human height has a 'middle' (the apex of the graph where the height of the largest percentage of the population is identified). The value of the 'middle' is called the mean. Other measures of 'middleness' are the median and the mode, but the mean is most usually used. It is useful to know the extent to which data is 'spread out' around a mean. The standard deviation is the most usual measure of 'spread-outness'.

Significant differences and drawing conclusions

When comparing means statistical tests help us to judge if the compared means are really different from one another or that any differences are just due to chance. Most biological studies set a confidence level of 95% or better that the means are really different from one another. If the results meet this confidence level, then we say that the means are significantly different.

> **Revision tip**
> **Variation**
> The term variation refers to the differences that exist between individuals of the same species, e.g. variations in the shapes of faces.

> **Revision tip**
> Figure 1 shows distribution curves of human height in two populations.
> The mean value of each distribution curve is the same. However the value of the standard deviation about each mean is different. The difference shows that the variation in height of population B is greater than in population A.

> **Revision tip**
> Remember that scientists assume that samples are representative of whatever it is they are investigating.

> **Revision tip**
> An investigation is designed to take samples at random. Random sampling means that any part of whatever it is being investigated, has an equal chance of being sampled.

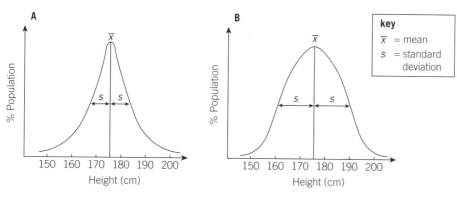

▲ Figure 1 *Distribution curves of human height in two populations*

Biodiversity

Worked example: Calculating standard deviation

The formula used to calculate standard deviation (SD) is:

$$SD = \sqrt{\frac{\Sigma(x-\bar{x})^2}{n-1}}$$

where

$\sqrt{}$ = square root

x = measured sample value

$\bar{x}$ = mean value of measured samples

n = total number of samples

Σ = sum of

Calculating standard deviation step-by-step runs as follows:

1. Calculate the mean of the observations (sample values) made.
2. Subtract the mean from each of the observations (sample values). Each result is a calculation of the deviation from the mean.
3. Square the deviations.
4. Sum (Σ) the squared deviations.
5. Divide the sum (Σ) by the number of observations (sample values) less one, i.e. $n - 1$.
6. Square root ($\sqrt{}$) the *dividend* (the result of the calculation) from step 5.

The result is a value for the standard deviation of the samples taken.

Summary questions

1. Distinguish between the mean, median, and mode of a data set. [3]

2. Explain why a characteristic showing continuous variation is likely to be polygenic in origin. [2]

3. Explain why the standard deviation is a measure of the variation of a characteristic. [2]

Chapter 10 Practice questions

1. The diagram represents a plan for organising living things into groups. Use the correct terms to fill in the blank spaces. Two spaces have been filled in for you. *(3 marks)*

2. Nature reserves are set up to conserve animals and plants. The effectiveness of a reserve is often determined by its size.

   a Small reserves can only support small populations. Explain why small reserves limit population size. *(2 marks)*

   b Three reserves of different size were established, each carrying a different number of a particular species of monkey. The amount of genetic variation in the monkeys of each reserve was measured after one, five, and ten generations of breeding. The results are shown in the table.

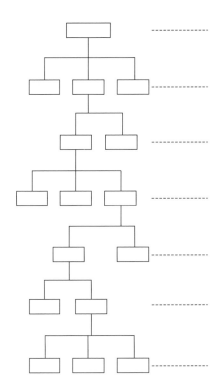

Number of monkeys in each reserve at the start	Percentage (%) of genetic variation after …		
	1 generation	5 generations	10 generations
2	75.0	24.0	6.0
10	95.0	77.0	60.0
100	99.5	97.5	95.0

   i Explain the differences in the % genetic variation in the monkeys of each reserve after 10 generations of breeding. *(3 marks)*

   ii Explain why the results suggest that a large reserve would better ensure the conservation of the species of monkey than a small one. *(3 marks)*

3. DNA extracted from the red blood cells of two different bird species was heated, separating its strands. Single DNA strands from the different species were then combined, giving a hybrid double strand. Differences between the nucleotide sequences of the different strands weaken the bond between them. The differences mean that the temperature at which the hybrid DNA dissociates is lower compared with pure DNA from either species. The difference in temperature at which the hybrid dissociated compared with the pure DNA is a measure of the genetic distance between the two species. The data can be used to estimate a date when the two species diverged from a common ancestor.

   The DNA of three different species of bird, A, B, and C, was extracted and hybrid DNA molecules prepared: AB and AC. Their respective dissociation temperature compared with the pure DNA of the different species was determined. The table shows the results.

DNA	Dissociation temperature (°C)
pure A, pure B, pure C	59.5
AB	56.3
AC	57.5

   a Explain how the data can be used to suggest the evolutionary relationships between the different species. *(5 marks)*

   b If 0.1°C difference in the dissociation temperature of hybrid DNA compared with pure DNA is equivalent to 1 million years before present (BP), suggest dates when species A, B, and C diverged from their common ancestor. *(2 marks)*

93

Answers to practice questions

Chapter 1

1. **a** Atoms of a molecule held together by sharing pairs of electrons in outer shells [1]
   b Positive / +ve charge at one position / H atoms of water molecule; negative / –ve charge at another position / O atom of water molecule [2]
   c Ions/polar molecules readily dissolve in water; form hydrogen bonds with other water molecules / polar molecules; high latent heat vaporization; cohesion tension; transpiration; heat capacity of water (biological significance should be paired with property of water as the result of its polarity) [3]

2. **a** Condensation reaction [1]
   b [7]

Category of biological molecule	Monomers	Chemical bond
Triglyceride	Glycerol; fatty acids	Ester
Protein	Amino acids	Peptide
Carbohydrate	Monosaccharides	Glycosidic

3. **a i** Rate of reaction increases; as temperature increases; increase temperature kinetic motion of substrate molecules increases; more substrate molecules strike active site; rate of reaction increases [4]
   ii Increasing denaturation of active site; alters shape of active site; fewer substrate molecules combine with active site; rate of reaction decreases [3]
   b Alters shape of active site; active site no longer complementary / matches shape of substrate molecule; substrate molecules cannot bind to active site [3]

4. Fewer peptide bonds, less concentrated protein solution; indicated by pink / violet colour; more peptide bonds; more concentrated protein solution; indicated by blue-violet/purple colour [3]

Chapter 2

1. **a** ATCG [Correct pair = 1 mark] [2]
   b X = hydrogen bond; Y = phosphodiester bond [2]
   c Hydrogen bonds [1]
   d DNA helicase [1]
   e S-phase [1]

2. **a** New DNA molecule consists of a strand of parent / original DNA molecule; strand that forms by complementary base pairing of nucleotides against parent / original strand [2]
   b One strand runs in 3'→ 5'→ 3' direction; partner strand runs in 5'→ 3'→ 5' direction
   (Accept description of 3'→ 5'→ 3' and 5'→ 3'→ 5') [2]
   c Nucleotides bind to template / original / parent strand; complementary base pairing; G binds C, A binds T [2]
   d i 28; arbitrary units [2]
   ii complementary base pairing; 44 arbitrary units of GC; (therefore) 56 arbitrary units of TA; divide (÷) by 2 / 56/2; 28 arbitrary units of T [5]

3. **a** X = adenine; Y = adenosine [2]
   b –OH group on C_2 of sugar (ribose); not –H atom on C_2 of sugar (deoxyribose) [2]
   c Phosphorylation [1]
   d Endothermic [1]
   e Light; oxidation of sugars / respiratory substrates (sugars / proteins / fats) [2]
   f Active transport, muscle contraction; activation during protein synthesis; synthesis of polymers / anabolic reactions (Any two of these or suitable other answers) [2]

Chapter 3

1. **a** X = intermembrane space; Y = crista / cristae; Z = matrix [3]
   b Ribosome. Translation of mRNA and protein synthesis. [2]
   c Attached or detached; depends on plane of section / angle at which cell cut; microtome; preparation of cells for electron microscope [2]

2. **a** Similarity, any one of: cell wall, cytoplasm, cell surface / plasma membrane; differences, flagellum; capsule; cell wall not made of cellulose [4]
   b Strands / loops of DNA / plasmids; not surrounded by (nuclear) membrane [2]
   c Glycan; chitin [2]
   d (Murein polysaccharide) cross-linked by peptide bonds [1]

3. **a** DNA replication during S-phase of interphase [2]
   b Strictly mitosis refers to division of nucleus; not division of cell [3]
   c i During metaphase chromosomes (as pairs of chromatids) align / line up on equator of cell; during anaphase pairs of chromatids separate, each chromatid of a pair moving to opposite poles / ends of cell [3]
   ii Centrioles [1]
   iii Spindle fibres not derived from centrioles [1]
   d i Checkpoints (help to) ensure mutated genes not inherited; mutation of genes regulating checkpoints; leads to loss of inhibition; cells with mutated gene(s) not prevented from entering next stage of cell cycle; leads to over stimulation; more cells produced than normal [5]
   ii New drugs target DNA replication; preventing DNA replication; new drugs target metaphase; interfere with spindle formation; metaphase cannot proceed to anaphase [5]

Answers to practice questions

Chapter 4

1. **a** Phosphate groups hydrophilic; phosphate groups face outwards; form hydrogen / H bonds with water molecules; hydrocarbon chains hydrophobic; chains point inwards; shielded from water (molecules); do not form hydrogen / H bonds with water (molecules) [2]

   b (Chain) X unsaturated / has double bond(s); (chain) Y saturated / only single bonds; double bonds kink hydrocarbon chains; phospholipids loosely packed; (cell) membrane fluid (idea of fluid mosaic model) [2]

   c Cholesterol molecules wedged in (phospholipid) bilayer; separating / spacing phospholipid molecules; maintains membrane flexibility, particularly at low temperatures [2]

2. **a** A: plasma / cell surface membrane pressed against cell wall; cytoplasm fills cell; B: plasma / cell surface membrane drawn / pulled away from cell wall; cytoplasm does not fill cell (Allow answer in terms of membrane or cytoplasm) [2]

   b A: water potential (WP) of cell more negative / lower than surroundings; water enters cell by osmosis; increase hydrostatic / water pressure presses plasma/ cell surface membrane against cell wall; B: WP of cell less negative / higher than surroundings; water leaves cell by osmosis; decrease hydrostatic / water pressure; cell surface /plasma membrane no longer pressed against / pulls away from cell wall [3]

   c ψ and ψ_s approx. equal value; if not then cell bursts; in absence cell wall constraining cell content [2]

   d ψ and ψ_s can have significantly different values; usually ψ_s greater than ψ; water enters cell by osmosis; ψ_p resists increase in hydrostatic pressure in cell; cell wall consist 40% cellulose; cellulose molecule high tensile strength; cell wall constrains cell content; plant cell does not burst; as hydrostatic pressure increases; cell becomes turgid when $\psi_p = \psi + \psi_s$ [3]

3. **a** Substance transported across membrane against its concentration gradient (Accept substance transported from high → low concentration) [1]

   b Shape / conformation of molecule of substance transported; matches / complementary to shape / conformation of carrier protein; protein shape is result of folding /coiling of protein molecule / its tertiary structure

   (Do not accept shape of transported molecule and carrier protein same) [2]

   c Similar shape molecules bind to carrier protein; binding sites occupied; molecules of substance normally transported cannot bind; cannot be transported [3]

Chapter 5

1. **a** Consist of more than one polypeptide chain [1]

   b Constant chain does not bind antigen; necessary for stability of antibody molecule; selection pressure against change in structure of molecule; variable region carries antigen binding site; shape / tertiary structure matches / complementary to shape of antigen molecule; body exposed to many / millions different antigens in a lifetime; selection pressure for many different forms of variable region [2]

   c Many different amino acid sequences of variable region; gives many different forms of tertiary structure of antigen binding site; particular antigen stimulates production of antibody with particular shape antigen binding site; matches / complementary to shape of antigen molecule; (Idea that complementarity between antigen binding site and antigen molecule basis of specificity) [3]

2. **a** Innate immune system: non-specific; immediate defence; acquired immune system: specific; long term defence [3]

   b Tetanus symptoms develop within 3–4 days; 2–3 weeks for tetanus antibodies to be produced; after exposure to antigen [2]

   c Absence of tetanus vaccination, no tetanus memory cells; anti-tetanus antibodies from another animal give immediate protection [2]

   d Passive immunity [1]

   e Antibodies produced by another animal are non-self antigens; stimulate immune response in person vaccinated; immune response destroys non-self antibodies / antigens [2]

   f Rapid division / clonal expansion of memory cells; gives secondary immune response quicker than primary immune response/ immune response to first-time infection; antigens destroyed before symptoms of disease develop [3]

3. **a** Genetic material is RNA; not DNA [2]

   b Binds to CD4 receptors; at surface of T helper (TH) cell; virus envelope / capsid fuses with cell surface / plasma membrane TH cell; viral RNA, reverse transcriptase / enzymes enter TH cell [1]

   c Destroys T helper cells; T helper cells affect activity of β cells and other T cells; two examples of effect e.g. produce cytokinesis / interferon/; promote proliferation β cells; increase production of antibodies [1]

   d Weakens immune system / responses (of HIV + person) less able to combat infections; infections opportunistic; infections develop which normally do not develop in healthy / HIV / –ve person [2]

Answers to practice questions

Chapter 6

1. **a** **i** A [1]
   ii B [1]
   b **i** Diffusion; concentration oxygen / O_2 in air in alveoli greater; than concentration oxygen / O_2 in blood; oxygen passes down oxygen / O_2 concentration gradient from alveoli to blood [2]
   ii 2 [1]
   c Flat / flattened [1]
   d Characteristic(s) of organism enabling it to survive in an environment [1]
   e Flat cells; diffusion pathway of oxygen / O_2 short; between alveoli and blood; rate of diffusion α 1/distance^2 [2]

2. **a** (Hard) exoskeleton contains chitin, impermeable to diffusion gases [2]
   b **i** X = spiracle
      Y = trachea
      Z = tracheole [3]
      ii Prevent collapse of tracheae; air pressure in tracheae less than atmospheric pressure [2]
   c Wing muscles contract vigorously when insect flying; oxygen / O_2 supply to muscle insufficient / not enough for aerobic respiration; muscles respire anaerobically; lactate / lactic acid produced [2]
   d Water potential more negative / lowered; compared with water potential ends of tracheoles [2]
   e More negative water potential muscle tissue; water passes from tracheoles to muscle; more air drawn into tracheoles; more air, more oxygen available [3]

Chapter 7

1. **a** A = right atrium; B = right ventricle; C = left atrium; D = left ventricle; E = vena cava; F = aorta; G = pulmonary artery; H = pulmonary vein [4]
   b Tricuspid and bicuspid valves closed; semi-lunar valves open [2]
   c Left ventricle pumps blood into aorta; thicker wall contractions more powerfully than right ventricle; aorta transports blood through heart → head / body circuit; greater distance than heart → lung circuit; blood supplied through contractions of right ventricle [3]

2. **a** 1 = plasma; 2 = tissue fluid; 3 = lymph; 4 = plasma [4]
   b At 1, hydrostatic pressure at arteriole end is greater; than oncotic / colloid osmotic pressure; hydrostatic pressure forces small molecules through walls of capillary vessels → tissues; net outflow of substances in solution / solutes from capillary vessels → tissues; at 4 hydrostatic pressure at venule end less than oncotic / colloid osmotic pressure; net movement of substances in solution / solutes from tissues → capillary vessels [5]

3. **a** **i/ii**

	Volume of blood (dm^3 min^{-1}) At rest	During strenuous exercise
Heart muscle	0.5	0.8
Skeletal muscle	2.2	2.0
Kidneys	1.4	0.7
Gut and liver	1.6	0.8
Skin	0.7	2.1
Brain	0.8	0.8

[6]

   b The rate of blood flow to these organs decreases; as blood diverted to skeletal muscle (and heart muscle) [2]

4. **a** Water vapour passes from inside of leaf to outside through stomata; majority of stomata perforate lower surface leaves; therefore blue cobalt chloride papers pressed against lower leaf surfaces change to pink more quickly than cobalt chloride papers pressed against upper leaf surface. [2]
   b Photosynthesis causes exchange of ions responsible for more negative water potential of guard cells compared with adjacent leaf surface (epidermal) cells; water passes from leaf surface (epidermal) cells down the water potential gradient into the guard cells; (their) increased turgor pressure of guard cells opens stomatal pore(s); water vapour passes from inside the leaf to outside; blue cobalt chloride paper turns pink; at night photosynthesis stops (absence of light); no exchange of ions; water potential gradient reversed; guard cells less turgid; stomatal pore(s) close; water does not escape from inside the leaf to outside; cobalt chloride papers remain blue (do not change to pink). [2]

Chapter 8

1. **a** Base of codon specifying position of amino acid residue in polypeptide chain; does not contribute to specifying positions of other amino acid residues [2]
   b If substitution results in alternative codon; encoding same amino acid; genetic code degenerate [2]
   c Peptide [1]

2. **a** **i** 6 [1]
   ii Genetic code is triplet; sequence 3 bases encodes position of amino acid residue in sequence of amino acids; triplet bases called codon [2]
   b UUACGUGAUAUGGGCCGG [1]
   c mRNA sequence is complementary to DNA sequence; T substituted by U/uracil in mRNA sequence [2]
   d AAUGCACUAUACCGGCC [1]

Answers to practice questions

    e Identical to DNA sequence, except that U substitutes T [2]

    f DNA sequence consists of exons and introns; in mRNA transcribed against DNA; introns edited / removed from mRNA transcription; forming mature mRNA [3]

3 Length double stranded DNA bound to proteins; proteins are histones; chromosome consists of DNA bound to histones; DNA replicates (S phase of interphases); each DNA replicate bound to histones forms chromatid; (after S-phase of interphase) chromosome consists two (pair) identical chromatids; joined by centromere; each chromatid (of pair) becomes (single) chromosome after telophase (mitosis) [5]

Chapter 9

1 a i A [2]

    ii B; (chart A shows all intermediate lengths between the extremes of length; no intermediate colours for the categories of colour in chart B) [2]

  b 30% [1]

  c A mutation of the gene(s) controlling colour of fish [1]

2 a The data is best represented as a bar chart because it is discontinuous [3]

  b Temperature; supply of nutrients in the soil; availability of water [2]

  c Seeds are result of sexual reproduction; so variation in offspring will be result of genetic causes; environmental cause of variation in fruit length of parent plants cannot be inherited (usually) [4]

3 a Discontinuous variation [1]

  b The frequency of distribution of the gene(s); which determines an individual's ability to taste phenylthiocarbamide; varies between different groups of people [3]

Chapter 10

1 a [Kingdom], phylum, class, order, [family], genus, species [3]

2 a Limited amounts of available food; and space; limit population size [2]

  b i Potential gene pool of the population of monkeys in each reserve is different; smaller the population, the less genetic variation there will be in subsequent generations; because of restricted number of mates; restriction in possible gene recombinations as result of mating; reduces genetic variation; compared with larger populations [3]

    ii Large nature reserve is able to support a large number of monkeys, greater genetic variation in the population; greater the number of monkeys, the greater the possible number of matings; and therefore possible recombination of genes; greater therefore is the genetic variation in the population; from generation to generation [3]

3 a The lower the hybrid DNA dissociation temperature; compared with the dissociation temperature of pure DNA; the greater is the genetic distance between species; species C is more nearly related to species A; than it is to species B; because difference in the dissociation temperature of the hybrid AC molecule compared with the pure DNA; is less than the hybrid AB molecule [5]

  b Species B diverged from its common ancestor with species A and C 32 million years BP; species C diverged from its common ancestor with species A 20 million BP [2]

Answers to summary questions

1.1

1. Ionic bonds carry charge, covalent bonds do not [1]
2. Polar molecules form hydrogen bonds with water molecules; therefore readily dissolve in water [1]
3. a. Large molecules consisting of many (thousands) carbon atoms joined together; polymers are macromolecules [2]
   b. Condensation reactions [1]

1.2

1. Ratio of hydrogen atoms to oxygen atoms is always 2:1 [1]
2. Have the same molecular formula ($C_6H_{12}O_6$); different structural formula [2]
3. Does not give a positive result in a simple Benedict's test; then add dilute hydrochloric acid to sample solution (acid hydrolysis); then add sodium hydrogen carbonate until the solution stops fizzing (neutralisation); then add Benedict's reagent and heat, result is positive [4]

1.3

1. Condensations: form chemical bonds between monomers; water molecule(s) produced. Hydrolyses: water breaks bonds between monomers. [2]
2. Acid hydrolysis of sucrose releases the monomers α-glucose and β-fructose; the monomers are reducing sugars that give a positive result when heated with Benedict's reagent.

1.4

1. α-Glucose: –OH group on C1 below the plane of the ring; β-glucose: –OH group on C1 above the plane of the ring [2]
2. Wall of plant cells 40% cellulose which has high tensile strength; contains content of cell (cell does not burst) when hydrostatic pressure within the cell increases (because of osmosis) [3]
3. Both molecules virtually insoluble; therefore do not add to solute concentration in cells; therefore water potential not affected (not made more negative); therefore osmotic movement of water into the cell unchanged [4]

1.5

1. Saturated: all chemical bonds of a hydrocarbon chain are single bonds *or* more H atoms cannot be added to the hydrocarbon chain; unsaturated: one *or* more chemical bonds of a hydrocarbon chain are double bonds or more hydrogen atoms can be added to the hydrocarbon chain [2]
2. Particles of substance are not dissolved but dispersed in a volume of water [2]

1.6

Go further

a. Consists of 3; polypeptide (chains)
b. Ligaments: attached to skeleton **across** joints; hold bones together at joint; elastic; (so) allow movement of bones at joints; tendons: attach muscles to bones; inelastic; (so) force generated by/of muscle contraction; transmitted to bone; moving bone
c. Polypeptide chains coiled into a triple helix; covalent hydrogen bonds hold polypeptide chains into triple helix; every third amino acid in polypeptide chains in glycine **Allow** glycine abundant/lot of; (so) allows polypeptide chains to tightly pack together; (so) adds strength to molecule; (side chains of) other amino acids hydrophobic; (so) molecule insoluble in water

Summary questions

1. C, H, N, O, sometimes S [2]
2. Add alkaline $CuSO_4$ solution (Biuret reagent) to sample; blue colour turning to pink/purple indicates presence of peptide bonds; therefore protein [3]
3. 2°: hydrogen bonds only; 3°: hydrogen bonds *and* ionic bonds, *and* van der Waals forces [3]

1.7

1. Amount of energy required to cause a chemical reaction; lowers activation energy [2]
2. Active site and substrate are fully complementary *only* after binding between active site and substrate occurs; initial binding of substrate molecule alters tertiary structure of active site making it fully complementary to substrate molecule [3]

1.8

1. The number of collisions of substrate molecules with the active site of enzyme increases with increase in temperature; more enzyme–substrate complexes form [2]
2. Addition of enzymes increases number of active sites of enzyme available to bind with substrate; rate of reaction increases; when all substrate molecules bound to active sites further addition of enzyme does not increase rate of reaction because no substrate molecules available to bind with active sites [4]
3. Denaturation: tertiary structure (shape) of active site permanently altered; deactivation: insufficient (heat) energy to enable significant number of collisions between substrate molecules and active sites [2]

Answers to summary questions

1.9

1. Competitive inhibitor combines with active site; combines with part of enzyme *other* than active site [2]
2. Non-competitive inhibitors *irreversibly* altering shape of active site [1]
3. Negative feedback [1]

2.1

1. Information (polypeptides/characteristics) in DNA/RNA (nucleic acids) is inherited (passed on to) daughter molecules of DNA/cells/offspring [1]
2. Phosphodiester bonds: join nucleotides forming polynucleotide chain; complementary base pairing: C bonds G, A bonds T (DNA) (A bonds U, C bonds G (RNA)); hydrogen bonds join complementary base pairs, therefore forming double-stranded molecule of DNA (RNA usually single stranded) [3]
3. Ribose sugar, deoxyribose (DNA)/ribose (RNA); phosphate group; base (A, T, C, or G (D N A), A, U, C, or G (RNA) [2]
4. mRNA: transcribed from DNA; tRNA: binds amino acid; brings amino acid to mRNA; rRNA: binds to protein; forming ribosome [4]

2.2

1. DNA helicase: catalyses breaking of hydrogen bonds between complementary bases; DNA polymerase: catalyses the formation of phosphodiester bonds between nucleotides forming a polynucleotide strand (strand of DNA/RNA) [2]
2. New daughter molecules each consist of a parent molecule strand of DNA which is the template enabling nucleotides each to bond with its complementary base on the parent strand; a new double-stranded molecule of DNA is formed consisting of a parent strand of DNA and a new strand of DNA joined through their complementary bases [2]
3. A strand of DNA (template) against which a new strand forms; its bases complementary to the template strand [2]

2.3

1. Light, sugars (other respiratory substrates) [2]
2. Adenosine consists of the base adenine joined to ribose (sugar); 3 phosphate groups are attached to the ribose (C_1) component of adenosine [2]
3. As chemical bonds break and form; more energy is taken in than released during the reaction [2]

2.4

1. a Charge is unevenly distributed through the water molecule; O atom carries negative charge, H atoms each carry positive charge; H atoms (positive) of a water molecule attract O atom (negative) of another water molecule [3]
   b Hydrogen bonds form holding water molecules together; enabling characteristics of water essential to life (cohesion/high heat capacity) [2]

2. Heat: thermal energy increases kinetic motion of molecules;
   temperature: measure of the motion of molecules;
   heat capacity: water can absorb more heat before temperature changes; heat breaks hydrogen bonds between water molecules instead of increasing their kinetic motion; (very) large number of hydrogen bonds broken down by heat, before water temperature rises; damps down large swings in temperature; provides stable environment [4]
3. Cohesive forces between water molecules high; enable unbroken columns of water; to move through xylem [1]

3.1

1. Magnifying power: the number of times larger an image is of an object seen in a microscope;
   resolving power: ability to distinguish between structures lying close together [2]
2. Wavelength of electrons (0.005 nm) is much shorter than visible light (400 – 700 nm) [1]
3. Each category of organelle has a particular density (mass unit vol^{-1}); each category spins down (settles at the bottom of the centrifuge tube) at a particular spin speed (g force/rpm) [2]

3.2

1. TEM: electrons pass through (transmitted) specimen and hit phosphorescent screen; screen glows on impact of electron; more electron hits more screen glows
   SEM: electrons scan surface of specimen and reflected; electrons captured; computer processes image [3]
2. TEM: magnets; light microscope: glass lenses [2]
3. An object observed in an image of a specimen seen in a microscope which is not a part of the specimen in real life; result of the preparation of specimen for microscopy [2]

3.3

1. $A = \dfrac{I}{M} = \dfrac{2.0}{100} = 0.02$ mm; divide by 4 to obtain length of 1 cell $= \dfrac{0.02}{4} = 0.005$ mm = 5 µm [3]
2. Able to obtain mean length of cell [1]

3.4

1. Enables different processes to occur in a cell at the same time [1]
2. a Where translation/synthesis of protein/polypeptide occurs [1]
   b Where energy is released by aerobic respiration of respiratory substrates (e.g. glucose) [1]
   c Rough: transports proteins synthesized on attached ribosomes towards the Golgi apparatus [1]
   d Packages different substances (e.g. carbohydrates and proteins forming glycoproteins); vesicles bud from Golgi apparatus transporting substances to cell surface membrane [1]

Answers to summary questions

3 Similarities: made of membranes; convert energy which is used to synthesise ATP; process of ATP synthesis fundamentally similar (chemiosmosis), proteins of electron transport chains part of membranes; arrangement of membranes increases surface area

Differences: chloroplasts larger than mitochondria (approximately x10 (one order of magnitude)); in chloroplasts thylakoid membranes stacked, in mitochondria inner mitochondrial membrane folded forming cristae; in chloroplasts space usually called stroma, in mitochondria usually called matrix [6]

3.5
Go further
a Strictly mitosis refers to division of nucleus (behaviour of chromosomes/chromatids); cytokinesis refers to cell division

b Gametes /sex cells/sperms and eggs; red blood cells (no nucleus)

c Carry out a particular/specific function; e.g. nerve cells (neurons) transmit nerve impulses /action potentials; muscle cells contract; goblet cells produce mucus

d Only some genes expressed in (embryonic) cells; genes expressed (active) in different ways /different times / different intensity; (so) different polypeptides / proteins synthesised / made; (these) determine type of cell

Summary questions
1 Key words: HOX genes; development genes [3]

2 Shape: flattened, specialisation maximising rate of diffusion; ciliated, moving material over cell surface (e.g. mucus); microvilli, increase surface area maximising rate of absorption/exchange of substances [3]

3 Tissue: group of similar cells with similar function; organ: group of tissues with a particular function; organ system: group of organs with a particular function [3]

3.6
1 Capsule; mesosomes; plasmids (but found in mitochondria, chloroplasts, yeast cells) [3]

2 Insert their DNA (RNA retroviruses) into host cell; virus DNA (virus RNA reverse transcribed to virus DNA) incorporated into DNA of host cell; viral DNA/mRNA translated into viral proteins; viral proteins assemble forming new virus particles; virus particles released when host cell dies [4]

3 Requirements for nutrients and water, and independent replication define life; viruses do not require nutrients and water, and cannot replicate independent of a host cell; viruses can be crystallised and stored dormant in the long term; addition of water reactivates them; bacteria show none of these characteristics [5]

3.7
1 DNA replication produces / results in two identical copies of genetic material / chromosomes; one copy passes to each daughter / new cell [3]

2 Plant cell wall firm / inelastic / not flexible; not able to form waist / constrict / pull inwards; cell plate forms in cytoplasm of dividing cell; extends outwards to cell surface membrane; splits parent cell into two daughter cells. [2]

3.8
1 Cell cycle describes formation of new (daughter) cells from old (parent) cells; checkpoints regulate progression of cells; through stages / phases of all cell cycle; genes control events at checkpoints; mutation of genes; control lost; cancers develop / over production of new cells / uncontrolled mitosis [3]

2 When mutagen / mutation does not result in cancer [1]

3 Proteins on cell surface membrane (of cancer cells); different (abnormal) from cell surface membrane proteins of healthy cells; abnormal cell surface proteins (of cancer cells) non-self antigens; immunotherapy facilitates immune system detecting non-self proteins; immune system mounts immune response; destroys cancerous cells [5]

4.1
1 Cell-surface membrane: separates cell contents from cell's external environment;

internal membranes: form organelles which compartmentalise cell enabling different processes within the cell to occur at the same time; the different proteins of membranes perform many of the functions of membranes (e.g. enzymes, transport) [3]

2 Fluid: because phospholipid molecules of bilayer loosely packed;

mosaic: scattering of protein molecules embedded in phospholipid bilayer [2]

4.2
Go further
If the temperature increases: phospholipid bilayer becomes more fluid/less rigid; because hydrocarbon chains/tails; of fatty acids; become more fluid/mobile/affects permeability of membrane; enables toxic/harmful substances to enter cell; causes damage/destroys cell; (may) denature proteins in phospholipid bilayer/membrane; affecting their function; give two examples e.g. transport of substances; action as enzymes; action as receptors

If temperature decreases: phospholipid bilayer becomes less fluid/more rigid; impairs cell movement; slows cell growth; reduces permeability; (therefore) reduces movement of substances into; out of; through cell

[At least **three** aspects of temperature increase **and** decrease should be considered]

Answers to summary questions

Summary questions

1 Because rate of diffusion $\propto 1/\text{distance}^2$, effective diffusion pathways are short; cell size limited because diffusion important mechanism for transport of substances across cell-surface membrane and within cell [2]

2 Protein complementary to (matches) the shape of the molecule; it passes across a membrane [2]

3 Difference in concentration of a substance between regions through which substance diffusing; divided by distance between regions [2]

4.3

1 Because water molecules pass from a region where they are in high concentration; down their concentration gradient; to a region where they are in lower concentration [2]

2 Osmosis occurs from a region of less negative (higher) water potential; to a region of more negative (lower) water potential [2]

3 The more solute in solution, the more its molecules restrict the kinetic motion of water molecules; few water molecules strike partially permeable membrane; fewer water molecules pass across partially permeable membrane [3]

4.4

1 Diffusion/facilitated diffusion are passive processes in the sense that the processes depend on the kinetic energy of the motion of molecules; active transport also depends on the energy released by hydrolysis of ATP, ATP → ADP + P_i [2]

2 Its hydrolysis ATP → ADP + P_i releases energy; that enables molecules of a substance to pass from where they are in low concentration to where they are in higher concentration [2]

4.5

1 Folds; villi; microvilli [2]

2 Co-transport of sugar (glucose) molecules **with/and** Na^+; facilitated diffusion of sugar/glucose; active transport of Na^+ [2]

5.1

1 Innate: non-specific, rapid/immediate response to antigens;

adaptive: specific, long-term response to antigens [2]

2 All the reactions of the body that make invading pathogens harmless [1]

3 Central tolerance develops during early embryonic development and development of B/T cells; cells develop surface receptors recognising self antigens; cells destroyed (clonal deletion) before maturity, preventing autoimmune response; peripheral tolerance develops as result of clonal deletion of late development T cells and mutated B cells; T cells carrying receptors recognising self-antigens unresponsive to them; clonal anergy [4]

5.2

1 Proteins bind to surface antigens facilitating recognition of phagocytes [1]

2 Move down concentration gradient of chemicals released by pathogens; chemotaxis [2]

3 Vacuole in phagocyte enclosing pathogens; lysosomes fuse with phagosome releasing enzymes which catalyse reactions destroying pathogens [2]

5.3

1 Respond to foreign antigens on surface of cells (antigen presenting cells); cell mediated response [2]

2 Promote proliferation of T cytotoxic cells and B cells; produce opsonins/cytokines; increase antibody production [3]

3 Perforin: binds to and perforates surface membrane of target cell; granzyme A poisons target cell (catalyses production of reactive oxygen species); granzyme B catalyses destruction of target cell [3]

5.4

1 B cells produce antibodies, T cells do not; B cells respond to foreign antigens in body fluids, T cells respond to foreign antigens on antigen presenting cells [2]

2 B cells divide in response to detection of foreign antigens [2]

3 Produced during first clonal expansion and persist; rapidly divide to produce plasma cells and more memory cells on exposure to same antigen that stimulated first clonal expansion; pathogen destroyed by rapid response before symptoms of disease develop [3]

4 Antigenic drift: minor mutations in pathogen resulting in new strains of pathogen;

antigenic shift: major mutations in pathogen resulting in new types of pathogen [2]

5.5

1 a Constant region: amino acid sequence of polypeptide chains similar in all antibodies

variable region: forms antigen binding site; amino acid sequence of polypeptide chains variable/unique to each type of antibody [2]

b Shape of variable region/antigen binding site complementary to shape of antigen that stimulated production of particular antibody; immune (antigen–antibody) complex forms [2]

2 Antibodies with same shape antigen binding site; complementary to same antigen; drugs bound to monoclonal antibody only delivered to diseased cells with antigens complementary to antigen binding site of monoclonal antibody; healthy cells not affected since do not carry antigens complementary to monoclonal antibody [3]

3 Agglutination, precipitation, act as markers, neutralise toxins, lysis [2]

Answers to summary questions

5.6

1. Preparation of dead/attenuated/harmless components of pathogens; antigens stimulate primary immune response in person receiving vaccine [2]

2. Active: person's immune system stimulated by antigens; memory cells produced; protection long term

    passive: result of vaccination with antibodies produced by another animal; memory cells not produced; protection short term [4]

3. Mass vaccination of people; population as a whole protected even though not everyone vaccinated; breaks chain of infection; 80–90% people vaccinated to achieve herd effect [3]

5.7

1. Genetic material is RNA not DNA [2]

2. Binds CD4 protein receptors on surface T_H cells; virus envelope fuses with surface membrane of T_H cell; viral RNA/reverse transcriptases enter T_H cell; reverse transcriptase catalyses viral RNA → viral DNA; viral DNA becomes part of T_H cell DNA; viral DNA transcribes/translates viral proteins; viral proteins assemble as new virus particles; particles released when T_H cell destroyed [4]

3. Any *three* of: receptor antagonists; fusion inhibitors; reverse transcriptase inhibitors; integrase inhibitors; protease inhibitors [3]

6.1

1. Larger the organism, the smaller SA/V; SA increases more slowly (power^2) than volume (power^3) [2]

2. As cell increases in size SA/V decreases; distance of diffusion pathways of substances increases; efficiency of diffusion processes decreases, division restores favourable SA/V ratio in smaller daughter cells [2]

3. Reduces body surface area to incident sun radiation / heat; (therefore) reduces warming; (therefore) reduces temperature rise [2]

6.2

1. Gases in solution diffuse down concentration gradients from where gases in high concentration to where in low concentration; concentration carbon dioxide high in *Amoeba*, lower in environment; concentration oxygen high in environment, lower in Amoeba; gases exchanged across surface membrane high → low concentrations [2]

2. Exoskeleton impermeable to gases in solution [1]

3. At rest: gases in solution fill ends of tracheoles; oxygen in higher concentration in solution compared with tissues; oxygen diffuses tracheoles → tissues; carbon dioxide in higher concentration in tissues than solution in ends of tracheoles, carbon dioxide diffuses tissues → tracheoles

    active: tissues respire anaerobically; lactate produced; water potential of tissues more negative; water passes by osmosis from ends of tracheoles into tissues; more air (therefore more oxygen) drawn into tracheoles; more oxygen available for exchange between tracheoles and tissues [5]

6.3

1. Many lamellae, each flat (large surface area) and thin (short diffusion pathway); each lamella folded into gill plates; blood flow through capillary vessels in lamella opposite to water flow over lamellae; counter current [2]

2. Concentration of oxygen in water greater than in blood; concentration of carbon dioxide in blood greater than in water; concentration gradients of gases between blood and water; counter-current effect; blood flow removes oxygenated blood from lamellae, concentration gradient oxygen between water and blood maximised; flow of water over lamellae removes carbon dioxide (in solution) from lamella surface; concentration gradient carbon dioxide between blood and water maximised;

    result: rate of gaseous exchange maximised [4]

3. Pharynx floor lowered; reduces water pressure in mouth cavity; water flows through open mouth into mouth cavity, high pressure → lower pressure; pharynx floor raised; increases water pressure in mouth cavity; pushes water over lamellae and against inside surfaces of opercula; opercula open [5]

6.4

1. Allow gases and water vapour to freely circulate within leaf [1]

2. Because concentration of oxygen used and carbon dioxide produced by aerobic respiration; is balanced by concentration of oxygen produced and carbon dioxide used by photosynthesis [2]

3. Loss of (evaporation of) water from inside leaf to the atmosphere; through stomata; water lost through transpiration replaced by water 'pulled up' through stem from roots; inside of leaf 100% humidity; concentration gradient of water between inside of leaf to atmosphere, high → low; transpiration maintained [4]

6.5

1. Opening/closing of stomata depends on turgidity of guard cells either side of stomata pore; when guard cells turgid internal hydrostatic pressure of each cell high; cells expand unevenly because inner wall of each cell thicker than rest of cell wall; uneven expansion opens stomatal pore; light and photosynthesis trigger the process; active exchange of ions (K^+/Cl^-) makes water potential of guard cells more negative (lower); water passes by osmosis from surrounding epithelial cells to guard cells; cells become turgid; in dark, photosynthesis and active transport of ions stops; ions diffuse down their concentration gradients out of guard cells; water potential of cells becomes less negative; water passes by osmosis from guard cells to surrounding epithelial cells, hydrostatic pressure in guard cells reduces; cells become flaccid; stomatal pore closes [4]

Answers to summary questions

2 Any characteristic increasing an organism's chances of surviving/reproductive success; stomata sunk in pits fringed by hair-like extensions of surrounding cells; branching root system maximising surface area available for absorption of water; leaves roll into tubes in dry conditions; stomata sunk in pits fringed by hair-like extensions of surrounding cells [5]

6.6

1 Alveolus wall thin (one cell thick); cells of alveolus wall squamous epithelium (flattened); walls moist (gases diffuse in solution); walls of capillary vessels supplying blood to alveoli one cell thick; cells of capillary wall squamous epithelium (flattened) [4]

2 Oxygen higher concentration in air in alveoli than in blood capillaries supplying blood to alveoli; oxygen diffuses from air in alveoli to blood in capillaries, high → low; carbon dioxide higher concentration in blood than in air in alveoli; carbon dioxide diffuses from blood to air in alveoli, high → low [4]

6.7

1 Inhalation: diaphragm/intercostal muscles contract; volume of thoracic (chest) cavity increases; air pressure in lungs reduced and less than atmosphere pressure; air passes into lungs from atmosphere high pressure → low pressure;

exhalation: diaphragm/intercostal muscles relax; volume thoracic (chest) cavity decreases; air pressure in lungs increases and more than atmospheric pressure; air passes out of lungs to atmosphere [4]

2 External intercostals: contract; raise ribcage; inhalation;

Internal intercostals: contract; lower ribcage, exhalation [3]

6.8

1 Vital capacity is total air capacity of lungs less residual volume; tidal volume is volume of air exchanged between lungs and atmosphere one cycle of inspiration/expiration [2]

2 Fibrosis: alveoli and associated capillary blood vessels twisted out of shape; lung tissue thickens and loses flexibility;

asthma: inflammation of the walls of trachea and bronchi (airways) narrows them; air flow restricted [3]

3 Destruction of alveoli reduces surface area available for gaseous exchange [1]

6.9

1 A carbohydrase is any of a number of enzymes (carbohydrases) catalysing digestion of carbohydrates; amylase is a particular carbohydrase catalysing digestion of starch → maltose [2]

2 Glycerol **and** fatty acids [2]

3 Endopeptidases: catalyse breaking of peptide bonds within polypeptide; exopeptidases: catalyse breaking of peptide bonds joining terminal (end) amino acids to rest of polypeptide [2]

6.10

Go further

a **i** Increase surface area (available for absorption)

  ii Cells: *microvilli*; surface: folds of *Kerkring*; *villi*

b Channel proteins; transport substances across surface membrane; specific to substance transported; fructose: facilitated diffusion; ATP not required/needed; down concentration gradient; glucose/amino acids: *co-transport;* sodium (Na^+) ions required/needed (form of active transport); ATP required/needed hydrolysis ATP ($\rightarrow ADP + P_i$) release energy required/needed again; again: concentration gradient

c (Triglycerides) combine with bile salts form *micelles;* movement of content of small intestine/peristalsis; delivers micelles to surfaces of cells lining villi/small intestine; micelles break down (at surface); release contents; contents absorbed by villus cells by diffusion; triglycerides resynthesised in villus cell; on smooth endoplasmic reticulum transferred to Golgi (apparatus); packaged within *chylomicrons*; pass to tissue fluid; by exocytosis; pass to lacteal vessels

Summary questions

1 Molecules co-transported/active transport with Na^+, through channel proteins within surface membrane of epithelial cell; ATP consumed/energy released; molecules pass from epithelial cell to blood in capillary vessel by facilitated diffusion; Na^+ actively transported from epithelial cell into tissue fluid; ATP consumed/energy released [4]

2 Combine with bile salts, forming micelles; micelles deliver their contents to epithelial cells of villi; at cell surface micelles release their contents which diffuses into villus cells; within cells triglycerides resynthesised on smooth endoplasmic reticulum; triglycerides transfer to Golgi apparatus; packaged with protein/cholesterol/phospholipid; packages called chylomicrons removed from villus cells by exocytosis to tissue fluid; enter lymphatic vessels; circulate in lymph; drain into subclavian vein via thoracic duct [6]

7.1

1 Refers to shape of molecule; consists of four polypeptide chains [2]

2 Part of protein; not made of amino acids [2]

3 Combines with oxygen molecule [1]

7.2

1 When haemoglobin is loading oxygen [2]

2 Different types of haemoglobin load oxygen at different partial pressures of oxygen depending on circumstances, e.g. haemoglobin of organisms living in oxygen depleted environment loads oxygen and becomes fully saturated at partial pressure of oxygen lower than haemoglobin of organism living in oxygen rich environment [2]

Answers to summary questions

7.3

1. Pulmonary arteries carry deoxygenated blood from heart to lungs; pulmonary veins carry oxygenated blood from lungs to heart [2]
2. Transports products of digestion from ileum (small intestine) to liver [1]
3. a Double: blood flows through heart twice; high pressure and flow rate oxygenated blood maintained to tissues; single: blood flows through heart once; loss of pressure and reduced flow rate of oxygenated blood to tissues [3]
   b High pressure/flow rate enables high metabolic rate and endothermy; low pressure/flow rate not able to sustain endothermy [3]

7.4

1. a Coronary arteries
   b Although filled with blood, heart wall too thick for efficient diffusion of oxygen, nutrients, and other substances throughout heart tissue [2]
2. Tricuspid/bicuspid valves closed; as blood fills atria, increasing hydrostatic pressure forces valves open; blood passes to ventricles; semi-lunar valves closed; as blood fills ventricles, increase in hydrostatic pressure; closes tricuspid/bicuspid valves; forces open semi-lunar valves; blood passes to pulmonary arteries (from right ventricle), to aorta (from left ventricle) [3]
3. Left ventricle pumps blood through heart → head/body circuit; greater distance than heart → lung circuit; thicker wall of left ventricle contracts more powerfully than wall of right ventricle; greater force propels blood greater distance [4]

7.5

1. Tricuspid/bicuspid valves open; semi-lunar valves closed [2]
2. Nerve impulses from sino-atrial node spread through walls of atria; atria contract; the nerve impulses stimulate atrioventricular node; nerve impulses from atrioventricular node pass along bundle of His consisting of Purkyne fibres; stimulate ventricles to contract; sino-atrial node called pacemaker; determines resting rate of heartbeat [5]

7.6

1. Differences include: arteries thick wall, veins thin wall; capillary wall one cell thick; veins have valves, arteries do not; blood high pressure in arteries, low pressure in veins, gradient of pressure high → low in capillaries [3]
2. Blood at low pressure; valves prevent backflow [2]
3. High blood pressure at arteriole end of capillary; net hydrostatic force high, forces out solutes; large molecules remain in blood; water potential of blood becomes more negative; water drawn into capillary; this oncotic force less than hydrostatic force low blood pressure and hydrostatic force at venule end of capillary; oncotic force greater; net force drawing tissue fluid into capillary vessel [6]

7.7/7.8

Go further

a i Pressure increases because of contraction of left atrium; opening the valve; blood flows from the left atrium to the left ventricle
  ii Pressure increases because of; contraction of left ventricle; opening the valves; blood flows from left ventricle into the aorta

b Wall of left ventricle thicker than right ventricle; contracts more powerfully; greater pressure; left ventricle pumps/circulates blood through head/body circuit (of blood vessels); longer than heart/lung circuit (of blood vessels); (through which) right ventricle pumps / circulates blood

Summary questions

1. Apoplastic: from cell wall to cell wall by capillarity
   symplastic: from cell to cell by osmosis [2]
2. Soil → root, osmosis; root → xylem, osmosis and capillarity; stem xylem → leaf, transpirational pull (root pressure also in non-woody plants); leaf → atmosphere, capillarity/osmosis [4]
3. Transpiration: refers to passage of water (and mineral salts) from roots to leaves/flowers/fruits
   translocation: refers to passage of nutrients from photosynthetic tissue to all parts of the plant [2]
4. Sources: where nutrients (sugars) are produced by photosynthesis
   sinks: where nutrients are consumed; differences in hydrostatic pressure propel nutrients from sources to sinks, high → low [2]
5. Water passes by osmosis from xylem to sieve tubes of phloem where water potential is more negative; because of loading of sugars from photosynthetic tissue into sieve tubes (source); increased hydrostatic pressure propels nutrients through sieve tubes; to tissues where sugars consumed; therefore water potential less negative and hydrostatic pressure reduced [4]

7.9

1. Aphids feed on sap; able to selectively locate sieve tubes, body detached from stylets inserted in sieve tubes; stylets exude sap, providing samples of sap [3]
2. Tree ringing removes bark containing phloem tissue but not xylem; after ringing trunk bulges above ring due to growth and accumulation of sugars; trunk does not bulge below ring [3]
3. ^{14}C injected into leaf; incorporated in sugars by photosynthesis; sugars transported in phloem tissue; sampling sap from phloem tissue (*see question 1*: aphids) enables ^{14}C to be detected; ^{14}C not detected in xylem [4]

Answers to summary questions

8.1

1. Base of a triplet specifying position of particular amino acid in polypeptide chain does not contribute to specifying positions of other amino acids [1]
2. Exons: lengths of DNA each encoding a polypeptide or part of polypeptide

   intron: non-coding lengths of DNA [2]
3. Short sequences of bases repeated a number of (many) times [1]
4. Sequence of symbols (letters/number) that encrypt (hide) meaningful information [1]

8.2

1. Pair of chromatids joined at the centromere [1]
2. A chromosome consists of DNA bound to proteins (histones); DNA replicates in the S-phase of interphase; the pair of chromatids represent a chromosome after replication; each chromatid is identical to its partner (sister) chromatid [2]
3. A loop of DNA; present in prokaryotic cells (and chloroplasts, mitochondria, yeast cells) [1]
4. DNA combined with histones (proteins); coiled forming nucleosomes; coils fold forming loops; loops coil (supercoiling) and pack together [3]

8.3

1. mRNA: single strand; tRNA: clover leaf shape; internal hydrogen bonding; has amino acid binding site; has anti-codon; CREDIT converse [2]
2. mRNA transcribed from DNA template; tRNA binds to RNA complementary codon; brings amino acid to ribosome [2]

8.4

1. DNA helicase: catalyses breaking of hydrogen bonds; between complementary bases; RNA polymerase: catalyses addition of RNA nucleotides to mRNA strand [2]
2. Ribozymes; catalyse own editing [2]

8.5

1. Transcription: length of mRNA forms against sense strand of length of DNA encoding protein (exon); mRNA is complement of exon; translation: mRNA attracts complementary tRNA/amino acid combinations; tRNA/amino acid combination binds to complementary codon of mRNA; peptide bonds join adjacent amino acids; polypeptide chain grows in length (elongation) [2]
2. A number of ribosomes which combine with mRNA; speeds up polypeptide synthesis [1]
3. Combination of amino acid with its particular tRNA requires energy released by hydrolysis of ATP; process called activation [2]

9.1

1. Mutation imvolving one (single) base of a codon [1]
2. A change of base of a codon that alters the codon to an alternative codon encoding the same amino acid; can occur because genetic code is degenerate [2]
3. Change in base sequence / codons; means encoding of amino acids changes [2]

9.2

1. Increases genetic variation through crossing over and independent segregation/assortment of chromosomes into gametes; reduction division (chromosome number halved in each gamete); prevents exponential increase (doubling) of genetic material from generation to generation [2]
2. Haploid gametes/daughter cells [1]
3. Gametes carry genes encoding variant characteristics; when gametes fuse (fertilisation) genes/variant characteristics inherited by (pass to) offspring that develop from zygote [2]

9.3

1. Total number of different alleles in a gene pool; represented by variation in genes of members of a population [2]
2. Gene pool refers to the total of all the alleles of all the genes of a population; gene flow refers to transfer of parental genes in a gene pool by sexually reproducing parents to offspring [2]

9.4

1. Stabilising selection: for any particular characteristic, extremes in variation of the characteristic are selected against; average form (e.g. body length) of characteristic selected for

   directional selection: average form and either of the extreme forms of characteristic selected against; other extreme form of characteristic selected for [3]
2. Change in organisms over/through time [1]
3. Mechanism of evolution; idea of competition [2]

10.1

1. Classification: grouping organisms (items) according to the characteristics the organisms share in common

   taxonomy: the methods and rules by which organisms are classified [2]
2. The characteristics shared by a group (taxon) of organisms are the result of evolution; stages in the evolution of the characteristics are the phylogeny of the characteristics and of the organisms identified with the characteristics; the characteristics define the taxon to which the group belongs [3]
3. Behaviour by one individual that initiates (releases) behaviour in another individual [2]
4. Enables individuals of same species to recognise one another; if different species, behaviour inappropriate; chances of sexual activity reduce; chance of production of zygote reduced [3]

Answers to summary questions

10.2

Go further

a % occurrence of genes/alleles; (in) population / gene pool / group of individuals of same species

b Alleles / gene favouring survival; selected for; increased survival; survivors more likely to (reproduce) pass favourable alleles / genes to next / subsequent generation(s); IDEA of characteristic(s) controlled by favourable alleles / genes enable individuals (with favourable alleles / genes) to adapt to changing environment; population evolves

c Differential survival; (of) individuals with favourable alleles / genes; controlling characteristics favouring survival; *more likely* to pass on favourable alleles to offspring

d Founder effect; genetic diversity of small population different; (from) larger population; (from) which founder population come / split from; IDEA of island populations; genetic bottleneck; large population reduced; few individuals survive; their subsequent population expansion; subsequent gene frequency of population different; original population

Summary questions

1 All of the species of a community of organisms; numbers of each species [2]

2 Species defined as group of interbreeding individuals producing fertile offspring; definition defines a population; definition means that a species shares a common gene pool and not the gene pool of another species; hybrids are offspring of parents of different species; many hybrids are fertile; hybrid parents and offspring share genes from gene pool of two (or more) different species; index of diversity not reliable because of question: 'what is a species?' [3]

10.3

1 Species providing resources not driven to extinction; if use of resource does not exceed capacity of species to reproduce themselves [2]

2 Gene pool of wild genetic material available to breed new varieties of crops/livestock from wild-type ancestors; essential should crops/livestock be susceptible to epidemic/pandemic outbreaks of disease [2]

3 Habitats are where species populations live in an ecosystem; conserve habitats and therefore conserve species populations; maintain biodiversity [3]

10.4

1 Easier to sequence bases than amino acids; may be more than one base sequence (codon) encoding particular amino acid (genetic code is redundant); fossil DNA more stable than protein [2]

2 mRNA is a complement of the DNA; against which the mRNA is transcribed [2]

3 Related species diverge from common ancestral species which would share a common gene pool; the more differences in the gene pool of each descendant species from the gene pool(s) of other descendant species, the more distantly related are the species; descendant species have fewer alleles in common with each other [3]

10.5

1 Mean: average value of a data set

mode: value that most often occurs in a data set

median: middle value of a data set arranged in descending order of value [3]

2 Because of range of intermediate values of characteristic between extreme values (high/low) of characteristic; single gene unlikely to encode intermediate values as well as average value of characteristic (e.g. human height) [2]

3 Standard deviation measures extent of spread of values of characteristic; either side of mean value of characteristic (e.g. human height: short → average → tall) [2]